Segmentation Analytics with SAS® Viya®

An Approach to Clustering and Visualization

Randall S. Collica

sas.com/books

The correct bibliographic citation for this manual is as follows: Collica, Randall S. 2021. *Segmentation Analytics with SAS® Viya®: An Approach to Clustering and Visualization*. Cary, NC: SAS Institute Inc.

Segmentation Analytics with SAS® Viya®: An Approach to Clustering and Visualization

Contents

About This Book

What Does This Book Cover?

The main purpose of this book is to demonstrate how to accomplish clustering and segmentation in SAS® Viya® through the use of several graphical user interfaces (SAS® Visual Statistics, Model Studio, SAS® Visual Analytics, and a coding interface of SAS® Studio). This is a "how to" book with practical, real data examples in each chapter. This book also covers visualizations of clustering including a relatively new technique called t-SNE that can be used to better understand the underlying structure of the data before and after clustering.

While this book does *not* cover the theory of clustering and segmentation, there are a good number of available references at the end of each chapter where you can find appropriate additional materials that are relevant to each chapter.

Is This Book for You?

This book is designed for analysts, data miners, and data scientists who need to use the all in-memory platform of SAS Viya for the purposes of clustering and segmentation. I have not attempted to write a comprehensive book on segmentation, but this book is focused on the analytics and methods of SAS Viya actions, procedures, and even the SAS 9.4 procedures that can be used in conjunction with SAS Viya through SAS Studio. If you are a novice with clustering and segmentation, a chapter in the Appendix is designed for the basic understanding of how clustering works with distances of observations. If you already know clustering well, then this book will aid you in how to accomplish clustering and segmentation analytics using SAS Viya.

What Are the Prerequisites for This Book?

While it will be helpful if you are already familiar with clustering concepts, it isn't mandatory as Appendix 2 provides a discussion on the basic concepts of distance and how that is used in many clustering algorithms. An understanding of basic analytics concepts such as linear regression, elementary probability, statistics, and machine learning will be helpful.

What Should You Know about the Examples?

Each example is real data that has been anonymized and is available for your use to understand how to apply each of the methods described in this book. The results that you obtain while executing each of the examples might differ from what is printed in this manuscript. The results might differ because of sampling proportions, distributed computing environments versus single computing, or other general options settings that can affect the results. One way to ensure that the results keep consistent in a distributed computing environment is to use the SAS Viya CAS node= restrictions that limit the processing to a single node. This will likely keep the results consistent; however, it may defeat the purpose of larger data sets where distributed computing is typically used to reduce execution times. Please see the "NWORKERS= Session Option" section in the *SAS® Cloud Analytic Services: User's Guide* concerning the CAS number of workers options available at: https://go.documentation. sas.com/?docsetId=casref&docsetTarget=n1v9x1q6ll09ypn0zknvo54rk9ya. htm&docsetVersion=v_001&locale=en.

Software Used to Develop the Book's Content

The software used in this book is SAS Viya version 3.5 and 4.0. (4.0 is the containerized deployment of SAS Viya and is labeled as 2020.1.x versions.)

Example Code and Data

This book includes tutorials for you to follow to gain hands-on experience with SAS. Each folder contains the data sets, code, and macros for each chapter except for Chapters 1 and 2. These chapters provide an introduction to clustering and segmentation with SAS and give you a flavor of the possible use cases. Chapters 3 through 6 all have examples, data, code where applicable, and any macros used in the code. You can access the example code and data for this book by linking to its author page at https://support.sas.com/collica.

We Want to Hear from You

SAS Press books are written *by* SAS Users *for* SAS Users. We welcome your participation in their development and your feedback on SAS Press books that you are using. Please visit sas.com/books to do the following:

- Sign up to review a book
- Recommend a topic
- Request information on how to become a SAS Press author
- Provide feedback on a book

About The Author

 **Randy Collica** is a Principal Pre-Sales Solutions Architect at SAS supporting the retail, communications, consumer, and media industries. His research interests include segmentation, clustering, ensemble models, missing data and imputation, Bayesian techniques, and text mining for use in business and customer intelligence. He has authored and coauthored eleven articles and two books. He holds a US patent titled "System and Method of Combining Segmentation Data." He received a BS in electronic engineering from Northern Arizona University.

Learn more about this author by visiting his author page at http://support.sas.com/collica. There you can download free book excerpts, access example code and data, read the latest reviews, get updates, and more.

Acknowledgments

I'd like to take this opportunity to acknowledge the many individuals who without their help and assistance this book would not have been possible. First, my developmental editor, Catherine Connolly, in SAS Press for her intense patience with me and my shortcomings in writing. Secondly, I'd like to thank Clare Casey, who tirelessly reviewed each chapter for accuracy, for her helpful comments and suggestions; this also goes for all the other technical reviewers inside and outside of SAS Institute who reviewed all or part of this manuscript. And last but certainly not least is my family—my very patient wife Nanci Collica and my children Janelle, Brian, Danae, and James. As Brian finishes his degree of MS of Applied Statistics at the University of California, Berkeley, where he will be entering the Data Science market shortly, we wish him all success in those endeavors.

Chapter 1: Introduction to Segmentation Using SAS

Introduction

In my previous book on segmentation and clustering (Collica 2017), I began with a description of what segmentation is, and I referred to Customer Relationship Management (CRM) in the context of marketing. CRM is the innate ability to understand as much about the customer in order to better serve the customer's needs, preferences, buying experience, and so on. SAS® Enterprise Miner™ and some SAS code were used throughout the book and now SAS offers a new addition to its analytics platform, namely SAS® Viya®.

SAS Viya is a high-performance computing architecture in which all computations are multi-threaded and contained entirely in-memory as each data set is contained in memory. In addition, the use of Massively Parallel Processing (MPP) processing is also possible as is using hardware acceleration using Graphics Processing Unit (GPU) devices. This allows much faster processing, and therefore algorithms that might have taken too long to be of practical use are now doable in a fraction of the time. This new architecture and the introduction of micro-services enables new algorithms and processes for data science while also allowing an open environment that supports SAS, Python, R, Lua, and Java. This book will describe how to use SAS Viya for clustering, segmentation, and visualization, while at the same time, introducing some new methods as well. My goal is to empower you with the knowledge of using SAS Viya for solving analytical business problems relating to clustering, segmentation, and visualization.

In the 1930s Chamberlin[2] had laid the foundation for segmentation by prioritizing the consumer over the producer by highlighting the significance of aligning products with the needs and desires of customers. Today really isn't too different except now the availability of different channels in which messaging and offers take place is much wider than ever before such as SMS text messages, telemarketing, direct mail, email, digital advertising, and personalized web offers.

Retailers can now install beacons in their brick-and-mortar stores. If a customer decides to accept the beacon service, the retailer using the beacon tracking system can know what aisle the customer is walking down in real time. For example, if the customer is in the shoe aisle, the retailer with pre-determined offers can then prompt an offer to the customer in real time! At check-out, that offer can be scanned at the register, and the marketing process can be a one-on-one communication by producer and consumer in real time at many locations simultaneously. This depiction is known as location-based marketing, and many organizations are aspiring to obtain this level of real-time marketing offer capability. But just the offer is only the start; the organization must have the infrastructure operating in order to capture the response, purchase, and record the transaction and preference directly into the customer data record. This is also true for the campaign management system's table schema for keeping accurate records and aligning specific offers and at times arbitrate among multiple offers.

The methods in this book will aid you in improving marketing effectiveness. By understanding who your customers are and what underlying homogeneous customer groups might be present within your data, you will have a greater insight about your customers. Armed with this new insight, you can more effectively develop offers to customers in specific segments that should have a higher propensity to purchase when such estimates are surfaced and acted upon. I am sometimes surprised at how often I find that customer segmentation efforts in industry are frequently lacking in the variety of data on their customer, patient, or account records. For example, if you only have web traffic data on your customers, then you are missing out on their demographics or firmographics if your customers are businesses. Likewise, if you have purchase transaction data and demographics or firmographics but you lack insights as to your customer's attitudinal information toward your organization, then again, you'll be lacking a complete 360-degree view of your customers.

Let's take an example of customer data where both behavioral and attitudinal attribute information exists on the same data set. A cluster segmentation of behavioral data is shown in Figure 1.1, which has seven clusters and some profile information. Figure 1.1 does not contain the attitudinal attributes in this model. However, in Figure 1.2 the attitudinal attribute was added with all other settings and attributes being the same from Figure 1.1.

Notice the changes to the channel_purchase attribute in the lower left quadrant (outlined in red) is vastly different in Figure 1.2 than in Figure 1.1. In this data set, a channel of "1" indicates purchase from a reseller only, a value of "2" is direct purchases only, and "3" is both direct and indirect reseller purchases. A value of "0" indicates neither and typically is return purchases. Adding just a single important attribute such as attitudinal information (values 1 to 5 as a nominal attribute) changes the cluster model significantly.

> **Key Message:** Behavioral and attitudinal data are typically much more influential in clustering for segmentation than demographic or firmographic data alone!

Figure 1.1: Cluster Segmentation without Attitudinal Data Attributes

Figure 1.2: Cluster Segmentation with Attitudinal Data Attributes

Segmentation: Art, Science, or Both

In keeping with the science of segmentation, may I digress a bit to review the other side of segmentation, namely the artistic or "left-brain" side? When planning for a segmentation project, whether it be for marketing, exploration, sales, customer, or patron similarities, or even patient likeness, the key to developing a segmentation that meets your organization's needs will undoubtedly involve your domain expertise and expertise from others as well. A keen understanding of the objectives of the segmentation project will greatly help in what approach you should undertake, what data you will need or perhaps even acquire, and the timeline needed to complete the project. In most segmentation projects, a typical outcome objective will be when the derived segments are relatively homogeneous within each segment and yet distinctly different from other segments. The following examples help to explain how both domain experience and a data driven method can be combined for a desired outcome.

This section briefly describes two use cases in segmentation that involved segmentation methods that were for strategic usage rather than a specific tactical purpose. One use case was for a product marketing analysis and the other for a sales segmentation focus. These two examples are from genuine business situations where product and sales executives needed guidance and direction that was based on data. In both cases, however, the businesses had some preconceived ideas about how to approach and recommend certain attributes based on their historical knowledge of the available data. Also, in both cases, the historical data had changed, and therefore, the data-driven approach that I took changed the course of the analytics that were originally based on those preconceived ideas. This is where data and business domain expertise must work together to achieve the desired goals and objectives needed by each of these organizational groups – hence both art and science.

Use Case 1: Strategic Product Segmentation

Late in 2009, a senior product line manager at my previous employer came to me and asked for my assistance in a project that required re-setting the course for this product line. At that time in the high-tech industry, changes in computer technology were migrating to more commodity-based hardware, and data centers around the US and the world were taking note and making changes in their purchasing decisions and plans for how data centers operate. Because of this apparent change in direction, the senior management needed some data-driven insights to assist them in choosing a course of action to take in the future. They desired to have data-driven assistance in guiding where the newer purchases were going in the market; however, they also had some preconceived ideas as to what the purchasing behavior might look like for certain high-volume purchase customers.

The senior manager had told me that there were four accounts that he believed I should focus on as my "target" purchasing trend behavior and on which I should fashion the strategic analytics accordingly. I immediately proceeded to extract the historical data for these four customer accounts to review the trends. However, the trend that the product line manager thought was a "good" trend turned out not quite what he expected. In fact, the trend was the opposite of

Figure 1.3: Four Account Trends Averaged

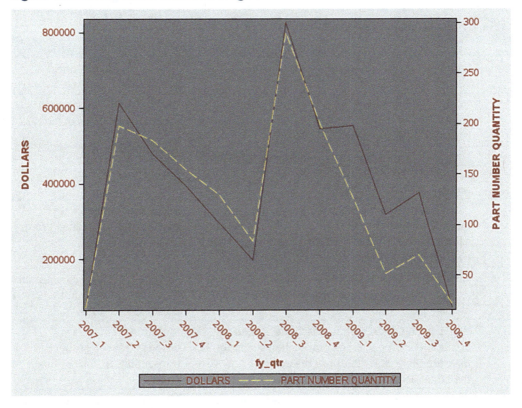

his initial expectation – it was heading downward not upward! Figure 1.3 shows the average of the four account trends (names omitted for proprietary reasons). As you can observe, these trends were not exactly what was expected. Although the volume in revenues and quantities was reasonably high, the trend certainly wasn't what we desired to set the pace for in the future. At this point of the analysis, I had a serious decision to make as to how to approach the strategic segmentation project.

The basic question I wrestled with was should I use the data as originally planned as the "target" trend with which to measure all other accounts and therefore segment the customer base, or should I use something different? I thought about this dilemma for a while and concluded that this would be a great learning moment for data-driven methods. I decided to keep the "target" trend. However, I would look for other trends that were better suited for the goals and objectives of the product line. The company chose these four accounts because they thought the trend was heading upward based on their experience with these strategic accounts. Indeed, the trend downward wasn't going to be helpful for this analysis. I set out to find accounts that exhibited a more positive trend even though their level of revenue and quantity wasn't quite near the level of these accounts. My rationale was that the *right* trend was more important to this project than the current volume of the four accounts. For this project I used SAS® Enterprise Guide® and SAS Enterprise Miner.

The steps I used in the data prep and analytics are outlined in Table 1.1 below.

Table 1.1: Steps Used to Prepare Data for Strategic Product Segmentation

Step	Process Step Description	Brief Rationale
1	Selected product lines needed in the products data table.	Query only product lines needed in this analysis.
2	Using the product line codes from previous query, I then queried the product purchase transactions between 2007 and the end of 2009.	Queried product transactions purchased on dates of interest.
3	I then aggregated the total revenues and quantities by purchase channel (direct or indirect) by customer account.	Aggregated customer transactions.
4	I then queried the transactions for the four accounts that the product manager desired. I also selected a few targeted accounts that had the product line transactions with increasing purchases over time.	Better product purchase transactions for a target group to measure against.
5	I labeled the accounts with the target transactions as well as the four originally selected by the product manager.	Account labeling accomplished by using a SAS format for unique account IDs.
6	With customer purchase transactions labeled by account ID in step 5, I ran a transaction similarity measurement[3] that measured the distance between the target accounts and all others.	Measuring the average distance from all product transactions to desired target group.
7	Merged the average transaction similarity metric along with the labeled accounts with the customer firmographic data and market share estimates using predictive models previously developed.	Final merge of data sets.
8	Performed clustering on the final data set and noted variables that affected the segmentation and profiled the segments.	Profiled cluster segmentation.
9	Generated charts and diagrams and gave presentation to senior product line managers with recommendations for next steps and direction.	Final analytic insights and recommendations.

The process in step 7 measures a target transaction against all other transactions in both the magnitude and time unit dimensions (Chamberlin 1933). The procedure gives a distance metric that when clustered together gives transactions of similar shape both in magnitude and in time sequence. This in effect performed transaction clustering along with customer clustering based on other firmographic information such as market share, industry group, and so on. All the data prep and queries were done using SAS code in SAS Enterprise Guide. The similarity metric used the SIMILARITY procedure (a SAS/ETS® procedure – SAS Institute 2014), and the cluster segmentation was done in SAS Enterprise Miner.

The clustering results produced a segmentation of 10 segments, of which three were considered high priority for future growth potential for the product directions. The variables that made an impact in this segmentation were first chosen as potential candidates based on my analyst's knowledge of the data, which I was very familiar with, and the estimated market share potential for which I previously developed analytic predictive models (Collica 2010). Table 1.2 and Figure 1.4 show a table and chart, respectively, of the variables that influenced the 10 clusters and their profiles.

> **Key Message:** Data-driven analysis that included the domain rules-based segment showed that the data-driven content was much more significant at driving the customer behavior segmentation than the rules-based segmentation.

Table 1.2: Key Variables Affecting Strategic Cluster Segmentation

Variable	Relative Importance
Company Segment	1.000
Industry Group	0.909
Log (ISS TAM)	0.780
RFM	0.778
Log (Similarity)	0.673
Log (Yrs Purch)	0.528
Orig Segment	0.321

Figure 1.4: Key Variables Affecting Strategic Cluster Segmentation

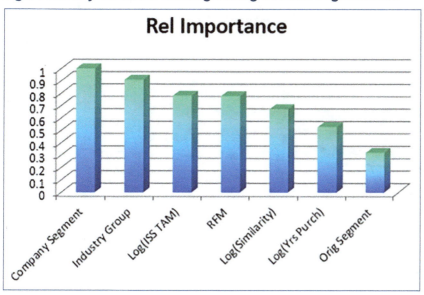

Figure 1.4 shows that Company Segment is the largest contributor, but the variable called Original Segment based on the manager's historical knowledge was the least important. The Company Segment variable was derived from the company's account reference file, which indicated whether the account was a corporate account, an enterprise or commercial account, or small and medium business (SMB) account, based on the definition that was applied as a business rule and was complied with by all countries. While the key message stated above is dominant for this analysis, it is not always true for all segmentation analytics. Of primary importance in this was the industry that the company was in and the estimated market called total addressable market (TAM). Recency, Frequency, Monetary value (RFM) is a typical segmentation that measures recency, frequency, and monetary value of purchases that is fully described in Chapter 4 of (Collica 2017). Once you have the estimated TAM at the customer account level, the estimates and other key attributes can be aggregated easily. The similarity metric also played an important factor. Figure 1.5 shows the value of the TAM per capita (total TAM divided by number of customers per segment) for the ISS product line group. Cluster segments 6–8 are considered the most valuable. Figure 1.5 shows the aggregate general relationship between the average similarity metric, ISS TAM, and the average number of years of customer purchase. This shows that there were non-linear relationships among these variables. The cluster segmentation did take this into account in the final clustering analysis. So the segments that show the highest average similarity and number of years purchase also had the highest average TAM.

The product-line management team was impressed by advanced analytical methods such as clustering of transactions into similar groups and using market share estimates. They also were aligned on the insights from the cluster segments that were of high-value and allowed more

Figure 1.5: Total Addressable Market Estimates by Strategic Cluster Segment

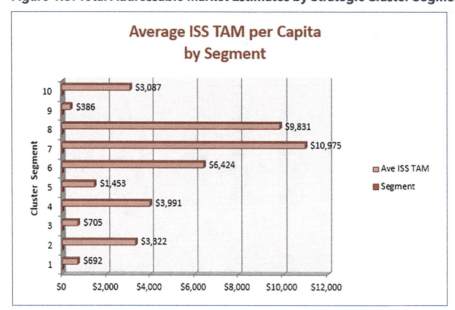

strategic plans to be developed. Again, what differentiated this from a tactical segmentation was the fact that I used semi-supervised techniques. A supervised technique requires a training data set where the correct answers are in the data, whereas an unsupervised technique has no labeled training data. Although clustering is an unsupervised method, I gave it some general direction by using variables such as a metric that measured how close the target transaction shape was to all other transaction shapes as well as an estimate of the market share. This market estimate was developed by me almost 11 years earlier using SAS Enterprise Miner with a two-stage model (Collica 2010). The product line management using the three high-value clusters and other more mediocre clusters was able to make definite strategic plans for customer accounts and target them much better according to industry and size, and so on. The elements that you place into the segmentation will strongly influence whether the segmentation can be used strategically or for tactical purposes. Market indicators such as share of wallet (SOW) help direct the segmentation for a strategic purpose rather than a tactical one.

> **Key Message:** The variables that you select for a segmentation will strongly influence whether the segmentation can be used strategically or tactically.

Figure 1.6: Plot Showing General Relationship between Ave ISS TAM, Similarity, and Yrs. Purchase

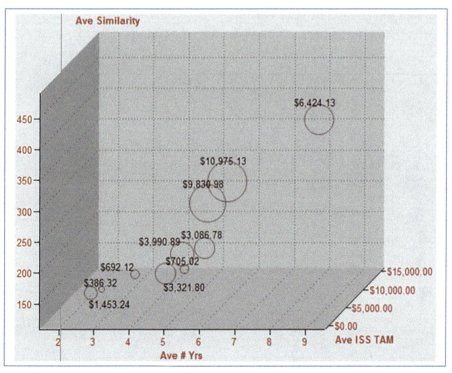

Use Case 2: Strategic Sales Segmentation

The key to this method is like Use Case 1 in that it requires an estimate of the revenue (or profit) share-of-wallet estimate in each account. When you can estimate the amount of total spending that the account can spend relating to the products and/or services that your organization can supply, the revenues that you generate from that account divided by the total estimated spending becomes the estimated share of wallet (SOW). This can be a very powerful metric if the estimate is reasonably accurate. Sales and marketing can use these estimates for strategic planning in areas such as the following:

- quota setting
- account prioritization
- product and messaging approaches
- segmenting customers by their SOW estimates

The business needed to segment the accounts so that their planning and goal setting process could be enhanced using the data-driven methodology to understand the potential versus their actual spending for IT products and services. The models that were developed in this SAS Global Forum paper were used in this particular business unit, and the planning process needed this segmentation and estimates at the account level, not just the total spending by industry like you can obtain from syndicated reports (Collica 2010). The plot in Figure 1.7 shows about 1,000

Figure 1.7: Account Revenues and SOW Percent

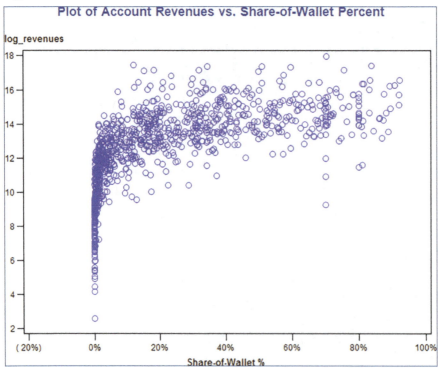

accounts with the logarithm of their latest year's revenues versus the estimated SOW percent. What this plot shows is the non-linear relationship between these metrics. However, it is very difficult to see any other pattern. One of the objectives that the sales management needed to accomplish by segmenting the sales accounts was to assist them in their sales planning and operations for the upcoming year. Historical segmentation methods relied heavily on a corporate segmentation based on historical revenues rather than current behavioral characteristics. Segmenting the accounts in Figure 1.7 with historical corporate segmentation methods didn't produce any usable analysis, so the segmentation proposed is to carve out some delineation of revenue and the SOW so that they could start the planning process. The segmentation in this case carved the SOW percent into three groups with splits at 10% and 40%. For the revenues, the two splits were $125,000 and $350,000. This produced nine segments. Now, when you fit a nonlinear model to the three different SOW levels, you obtain the plot shown in Figure 1.8.

Figure 1.8: Non-Linear Model Fit for Three Segment Levels

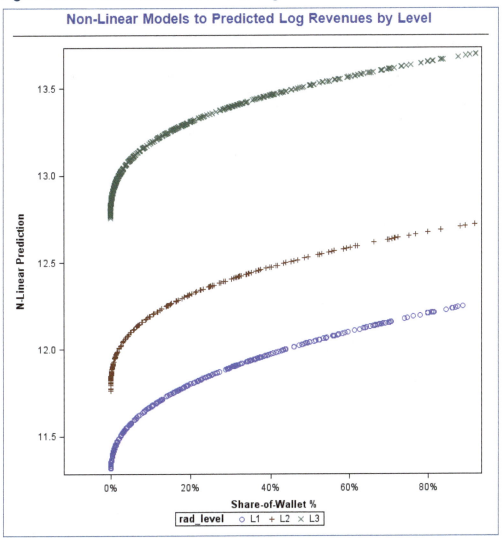

In Figure 1.8, you can more clearly observe the relationship for the different SOW segments. Each of the nine segments had a sales strategy that fit their SOW level and the total revenues that could be expected. Using other models in tactical campaigns, you could offer certain cross-sell/upsell products or services that better fit according to the expectations. The levels of R, A, and D represent Retain, Acquisition, and Develop. This strategy was used successfully in several of the major business units. This segmentation was also combined with other segmentations around the organization to improve the efficacy of marketing and could potentially be used in market research.

In the following chapters, you will learn how to use SAS Viya and SAS 9.4 together as one platform along with open source so that data scientists and analysts alike can use, collaborate, and deploy models in their organizations to gain insight and improve their businesses.

References

Chamberlin Edward. 1933. The Theory of Monopolistic Competition. Cambridge, Massachusetts: Harvard University Press.

Collica, Randy. 2010. "Estimating Potential IT Demand from Top to Bottom," Proceedings of the SAS Global Forum 2010 Conference. Cary, NC: SAS Institute Inc. Paper no. 372.

Collica, Randall. S. 2017. Customer Segmentation and Clustering Using SAS® Enterprise Miner™, 3rd ed., Cary, NC.: SAS Institute Inc.

SAS Institute Inc. 2014. Chapter 24, "The Similarity Procedure," SAS/ETS 13.2 User's Guide, Cary, NC.

Chapter 2: Why Classification and Segmentation Are Important in Business and Science

Some Applications of Clustering and Segmentation

If you were to make a list of the uses of clustering and segmentation, you would have a fairly long list of applications. A few are shown in Table 2.1 below.

Table 2.1: Some Applications of Clustering and Segmentation

Industry or Domain	Application
Science: Astronomy	Clustering of astronomical data to uncover new insights of our galaxy and universe as in the H-R diagram for classification of stars
Health care: Diagnosis	Clustering and segmentation of medical imaging (x-ray, magnetic resonance, and so on) for detection of potential cancer areas
Agriculture: Forestry	Clustering of tree types from images to detect changes in soil compositions
Retail: Customer/prospect data	Understanding of customers or prospects that have similar attributes for the purposes of direct marketing and loyalty initiatives
	Retail store clustering to inform merchandising and assortment localization
Telecom: Network sensor data	Clustering and segmentation of sensor metrics for early warning pre-failure signatures – anomaly detection
Financial: Time series of financial markets	Clustering of financial times series into groups of similar time-series patterns allows for further modeling of each segment that has relatively similar pattern metrics
Public sector: Segmentation of services	Use of data-driven segmentation to help transform government services, criminal justice reform, and transportation

Clustering is an unsupervised technique that helps in the overall process of having a segmentation where each segment has similar unique characteristics different from other segments. While clustering is one of the most common technique used in segmentation, it certainly isn't the only one. As we'll see later in Chapter 3, some clustering algorithms can be model-based methods that fall into the category of semi-supervised methods. If you were to query the web for the top 10 most popular algorithms used in data mining and data science, you'll find that k-means clustering still makes that list as of late in 2019 (upGrad 2019). I had done this same type of internet query back in 2006 and the k-Means clustering made that list then. So, if clustering using k-means is still in the top 10 list, the question becomes why is clustering so popular for such a length of time? I think the answer to that question lies in the use of clustering as a method for grouping customers, patients, and many other varieties of data sets into segments that have relatively similar characteristics in each segment. Profiling a data set is also another reason why clustering is often used so that a better understanding of the composition of underlying patterns and make-up of data sets. When I think of profiling, the term or phrase I give is "data assay," which comes from the term "bioassay," which means the measurement of the concentration or potency of a substance by its effect on living cells or tissues.

Table 2.1 is a rather short list, but the list could be enumerated to a great extent. The flavor of the varieties of applications shown should give you a feel for the types of segmentation uses. In the marketing context, the holy grail is to be able to market to segment of one; each person has a customized messaging and offer particular to their needs or desires. Conversely, if many customers get the same offer and message (mass marketing), then many customers or prospects might not accept the offer either due to the wrong message, wrong offer, or perhaps just wrong timing. Therefore, somewhere in between these two marketing extremes is what is needed. However, finding the right balance between what your organization can offer and the messaging and the number of customers or patrons you have in each segment is key. In the age of omni-channel marketing where several distribution channels currently exist, for example, web, mobile, SMS text message, direct mail, email, and social, the marketer needs to consider many attributes such as budget, channel usage by customer preference, how to optimize spending for maximum response, and return on investment. These are several of the considerations that a typical marketer needs to think about.

Segments of your customers enable you to understand the patterns that differentiate your customer base. However, just understanding how your customers are alike and different is not enough unless you are able to do something tangible with this new insight and understanding. With customer segmentation you should be able to customize the following:

- Identify the most and least profitable customer groups
- Build more loyal relationships
- Price product differently
- Deliver more personalized promotions and offers

It might be useful to think of customer data attributes as a strand of DNA or fingerprint for each customer. Figure 2.1 shows some information that is collected on each customer. It is sometimes

Figure 2.1: Examples of Customer Attributes as their "DNA"

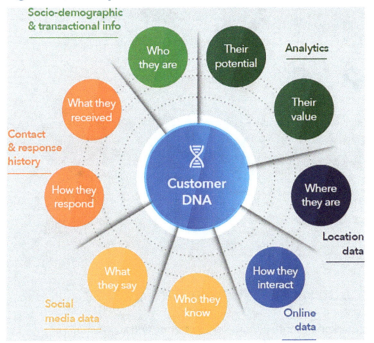

surprising that many organizations have very general segmentations, have little tailoring, and mostly a generic approach in their marketing efforts. Figure 2.1 is an example that illustrates this concept. There is a big difference in a telecommunications company presenting a customer with a random offer versus a specific data enhancement just before his or her data usage is running out while watching a movie on their tablet or smartphone. Having the right offer to the appropriate individual at the right time means much greater offer acceptance rates and likewise cross and upsell revenues!

Customer segments can come in a wide variety of uses such as lifetime value segments, revenue risk segments, attitudinal segments, purchase behavior segments, and customer journey segments, among others. A true data-driven customer journey approach example might look like the one in Figure 2.2.

In each phase of the customer journey, some set of attributes were collected over time and recorded and along with the organization's response to each journey segment. This segmentation journey can be very useful for understanding what a customer looks like when they first purchase or signs up for your service. This information might be used to estimate which set of prospective customers looks a lot like your existing customers, especially the high-value ones.

Figure 2.2: An Example of a Customer Journey Maturity Map

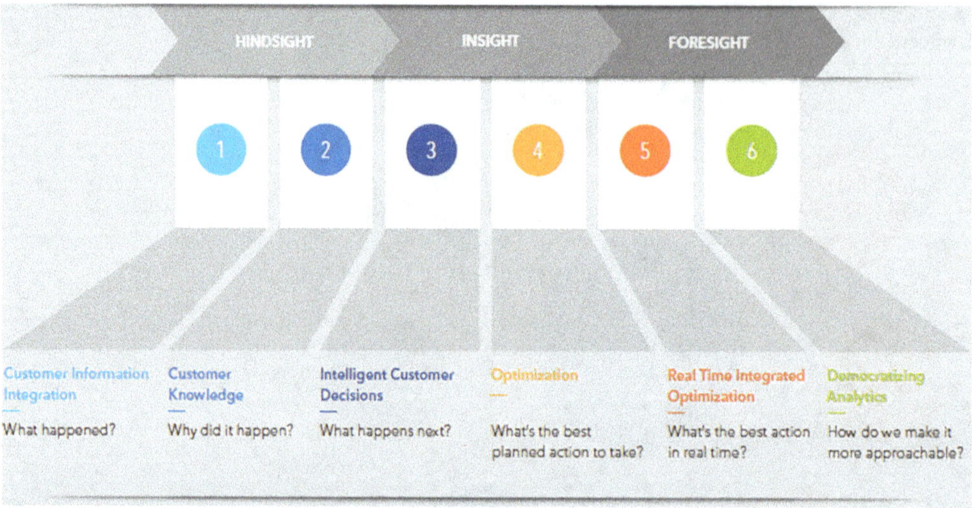

Clustering Applications on Unstructured Data

Clustering is often used in the marketing context; however, clustering and segmentation applications can help in understanding how health symptoms and outcomes might be linked in segmented patient groups that share similar characteristics. In the example below, we will review how we can analyze a data set that has patient symptoms, some vaccine information, and a few demographics to help determine whether there are certain life-threatening conditions that might have some common symptom characteristics among patient segments. When you know which part of the customer journey you're looking to understand, you can more easily identify:

1. The appropriate data attributes required.
2. How to affect change and improve the customer experience *at this point in the journey.*

The Vaccine Adverse Event Reporting System (VAERS) is a public, self-reporting data set of a sample of clinics and hospitals around the U.S. available at https://vaers.hhs.gov/data.html. This database is co-managed by the Centers for Disease Control and Prevention (CDC) and the U.S. Food and Drug Administration (FDA). VAERS is not designed to determine whether a vaccine caused a health problem, but it is especially useful for detecting unusual or unexpected patterns of adverse event reporting that might indicate a possible safety problem with a vaccine. The data set in this next example was taken from the VAERS data for 2016 and 2017. The objectives of the VAERS as stated on their website are:

- Detect new, unusual, or rare vaccine adverse events
- Monitor increases in known adverse events
- Identify potential patient risk factors for certain types of adverse events
- Assess the safety of newly licensed vaccines

- Determine and address possible reporting clusters (for example, suspected localized [temporally or geographically] or product-/batch-/lot-specific adverse event reporting)
- Recognize persistent safe-use problems and administration errors
- Provide a national safety monitoring system that extends to the entire general population for response to public health emergencies, such as a large-scale pandemic influenza vaccination program

The objectives in the following example are much simpler than those of the VAERS. This example uses:

- SAS© Visual Text Analytics for natural language processing
- SAS© Visual Data Mining and Machine Learning for data pre-processing and text

The example outlined in Table 2.2 uses SAS© Natural Language Processing (NLP) SAS© Visual Text Analytics, SAS© Visual Analytics, and SAS® Clustering in the Model Studio GUI interface all within SAS Viya.

> **The key objective** is to demonstrate how one can cluster unstructured textual data by determining key topics and then reviewing the profile of each segment to observe any similarities within the segments.

In Table 2.2, the steps needed to complete the work of this example are outlined as well as a brief rationale for each step. The detailed steps and explanations will follow. I will ou\tline more detailed steps in a later chapter for this exercise, but this should be helpful for how clustering can be used in health care.

Table 2.2: VAERS NLP Text Analytics and Segmentation Example

Step Number	Brief Process Step Description	Brief Rationale
1	Load desired data set into active memory.	Make data available for processing into memory.
2	Assign a target variable: L_Threat	Although we are doing a cluster segmentation flow, we still need a target variable assigned.
3	Add an Imputation node to impute missing levels.	Age variable has missing levels as does number of days in the hospital.
4	Add a Clustering node to determine a cluster segmentation.	Use range std. method for numeric inputs and relative freq. for class inputs. For the number of clusters – select Aligned box criterion and the estimation criterion is Global peak and PCA for the alignment method.
5	Add a Segment Profile node.	Use all default settings.

The results of the cluster segmentation enable us to find similarities of certain attributes in seven distinct segments. Figure 2.3a shows the pipeline process flow diagram, and Figure 2.3b shows some of the output results. Each node in the process flow has unique settings that the user can select. Seven segments were selected rather than automatically determining the number of potential segments. This was determined by trial and error and using some domain knowledge of the number of unique symptom text topics and the unstructured data in this data set. The automated clustering methods came up with five or 13; however, I realized the actual number is likely less than 10 but more than two or three. The Imputation node imputed the median value of two attributes: the age of the patient and the number of days since onset of the vaccination date (onset date – vaccination date). Figure 2.3b shows the Cluster node display output, and Figure 2.3c is a partial display output of the Segment Profile node's results. In the bottom left bar chart, the Variable Worth by Segment plot is shown for Cluster 1. While this is a simple example, it does indicate that profiling what the clustering algorithm found in each cluster segment is necessary and vital for key insights and further analytics to be useful to your business or organization. This example shows how NLP of unstructured textual data and machine learning (ML) can be combined to produce an output of cluster segments profiled in Figures 2.3b and 2.3c. While there are many ways one can profile clustered data, the SAS Visual Analytics reporting allows a flexible design with many object types to select from in forming an interactive report. Figures 2.4a-d show a snapshot of each of the four tabs in the report. While the analytics reporting will be detailed in a later chapter, this example gives you some ideas on how you might want to display the information of the segmentation for the users.

Figure 2.3a: Pipeline Process Flow Diagram of Clustering

(Continued)

Figure 2.3b: Partial Output Display of Cluster Node

Figure 2.3c: Segment Profile Node Results

Figure 2.4a is a high-level overview where each cluster segment can be highlighted and the hospital days heat map and histogram are sub-set on the cluster ID selected. The second tab in Figures 2.4b and 2.4c is a cluster dashboard showing a cluster frequency count of one of the key topics, and the dial indicators show the normalized cluster distance grouped by level of threat (Y/N). The fourth and last tab in Figure 2.4d shows an interactive logistic regression model with the left bar chart showing the fit summary of each attribute and its p-value assessment.

> **Key Message:** Clustering and segmentation can take many forms depending on the business objectives. The key to the analytics behind clustering and segmentation is to find the best methods that suit the business needs.

Another example of segmentation profiling is the use of Microsoft Excel. Summary statistics used in Figure 2.4c can be exported from SAS Viya to a CSV file and easily imported into Excel. Figure 2.5

Figure 2.4a: Sample Profile of Cluster Segmentation Tab1

Figure 2.4b: Sample Profile of Cluster Segmentation Tab2 (continued)

(*Continued*)

Figure 2.4c: Sample Profile of Cluster Segmentation Tab3 (continued)

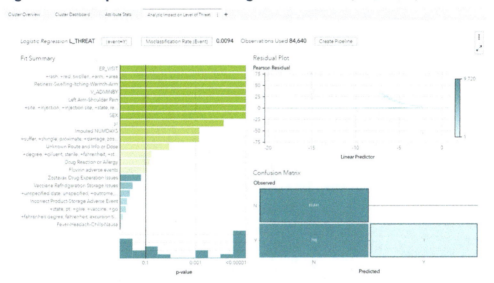

Figure 2.4d: Sample Profile of Cluster Segmentation Tab4 (continued)

Figure 2.5: Strategic Segmentation Profile Example in Microsoft Excel

Cluster Segment	Cluster Name	Brief Segment Description	# in Cluster	Per Capita ISS TAM	Total ISS TAM ($M)	Ave Scale-Out Similarity	Ave # Yrs Purchase HP
1	SMB Territory1	SMB Territory Accounts, Tart. Industries (Utilities, Prof. & Bus. Svcs, Construction, etc.).	23,003	$ 692.12	$ 15.92	188.27	4.01
2	SMB Commercial	SMB & Commercial Accounts, 50% MFG - 20% FIN - 15% HLS, Medium TAM.	6,812	$ 3,321.80	$ 22.63	174.56	4.60
3	SMB Territory2	60% SMB Territory, 26% Small SMB, 15% Commercial Accounts, Tart Industries, Low Scale-out potential smallest total ISS TAM.	4,306	$ 705.02	$ 3.04	197.58	5.40
4	The Fed	Pub Sector (Fed,Loc) Accounts. Med Scale-out potential, moderate total ISS TAM.	7,156	$ 3,990.89	$ 28.56	201.63	5.00
5	The Manuf	Manuf. Cluster, 60% Corp - 22% Ent - 17% Comm with some Edu. Accounts. Low Scale-out potential youngest HP customers, moderate per capita ISS TAM.	3,357	$ 1,453.24	$ 4.88	153.43	2.67
6	High-Value Mix Bag	55% SMB, 36% Commercial Accounts, Mixed Industries, Longest HP customers. High Scale-out potential. Largest total ISS TAM.	9,693	$ 6,424.13	$ 62.27	405.25	8.63
7	CME/HLS	CME and HLS Segment, Higheset Per Capita ISS TAM, Med Scale-out potential. Split in Corp/Ent/Comm Accounts, largest group of original scale-out customers.	2,139	$ 10,975.13	$ 23.48	278.31	5.59
8	Finance Group	Financial Segment, Corp/Ent split, 2nd largest per capita ISS TAM, Med Scale-out potential, smallest number of customer sites.	1,427	$ 9,830.98	$ 14.03	249.25	5.15
9	Small SMB	100% small SMB with no Amid Accounts, smallest per capita ISS TAM, 2nd youngest HP customers.	23,193	$ 386.32	$ 8.96	166.07	3.09
10	Education	Mostly Edu Accounts, Med Scale-out potential for High-Ed, Med-High total ISS TAM and Med Scale-out potential for High-Ed customers.	11,092	$ 3,086.78	$ 34.24	217.39	5.75

is an example of such a segmentation project I did at a previous employer for a product division where a strategic segmentation was used to help determine potential market areas of opportunity.

As you have seen some brief examples of clustering and segmentation in different business settings, I believe the key answer to why segmentation is important to many organizations is that it allows profiling of large data sets by finding similarities among the data observations using the attributes in that data set and grouping those similarities into segments. Each segment then has several unique features that describe and profile what is most common among the observations within that segment. In short, clustering and segmentation is one way of discovering more insights about your data and thus your customers, patients, or whatever set of observations is important to your organization.

References

upGrad. 2019. "Top 10 Most Common Data Mining Algorithms You Should Know," upGrad blog, published Dec. 2, 2019, available at https://www.upgrad.com/blog/common-data-mining-algorithms/.

Chapter 3: The Basics of Clustering and Segmentation in SAS Viya

Introduction

SAS offers several ways that you can cluster and segment your data using SAS Viya. We'll start off with very simple, quick, and easy methods to use and get a little more sophisticated in each subsequent example. The same data set will be used so that you can compare each approach a little easier. The data set that we will use is called CUSTOMERS, and it is business-to-business retail purchases of IT hardware equipment on about 100,000 customers. The outline of techniques in each task will be as follows in Table 3.1.

Table 3.1: Chapter Exercises and Examples

Task	Description
1	Using SAS Visual Statistics for clustering
2	Using SAS VDMML (Visual Data Mining and Machine Learning) in Model Studio
3	Using programming in SAS Studio with CAS and procedures

Task 1: Using SAS Visual Statistics for Clustering

Let's jump right in and get started. Table 3.2 outlines the steps and rationale for each task in the process. Then each step will be outlined in more detail. The version of SAS Viya being used in these examples is 3.5.

Table 3.2: Clustering Customers Data in SAS Visual Statistics – Task 1

Step Number	Brief Process Step Description	Brief Rationale
1	Load the CUSTOMERS data set into active memory.	Make data available for computations and analysis.
2	In the **Analytics Life Cycle** menu select **Explore and Visualize**.	Start SAS Visual Analytics for visualization and exploration of the CUSTOMERS data set.
3	Since we are exploring rather than reporting, select the **Start with Data** icon.	Review the data attributes of the CUSTOMERS data set.
4	In the available list of data sets, select the **CUSTOMERS** data set for use.	Run through a brief profile of CUSTOMERS data set.
5	Select **Profile** and run the Profile.	Observe number of missing observations and so on.
6	After you have reviewed the profile, select the **OK** button.	Select the blue OK button. Data is placed in Report1 template.
7	In the **Measures** listing, right mouse-click on **0-No, 1=Yes** to convert to a Category. Then drag this public sector attribute onto **Report1** on the right.	Explore the Public Sector yes/no attribute as a bar chart item.
8	Click the + next to Page1 to open a new page.	Develop a second page to the analytics report.
9	Click the **Objects** icon item at the far left of your browser page.	Bring up a long list of potential objects that you can add to your new page design.
10	Scroll down the Object items until you reach the **Visual Statistics** group. Select the **Cluster** object and drag it onto **Page2** on the right.	Add a Cluster object to perform distance-based clustering on the training portion of the CUSTOMERS data set.
11	Now to the right of the Cluster object click the **Roles** icon. Click the + to add variables to cluster: Add **Corp Revenue last fiscal yr., Est. Product-Service Spend, No of Local employees**, and **No of Years Purchase**.	View interactive clustering object on a few attributes.
12	Figure 3.3 should show what your clustering object result should look like.	Visual Statistics Clustering output review.

(Continued)

Table 3.2: Clustering Customers Data in SAS Visual Statistics – Task 1 (*Continued*)

Step Number	Brief Process Step Description	Brief Rationale
13	Distribution of revenues. Start of transforms. Click the **Data** icon at the far left and then click the **+New** data item.	Review how the attributes in the clustering are distributed. They might need to be transformed for better results.
14	Select the **Calculated item** icon and then open **Numeric**. Select the **Operators** field and choose **Numeric (advanced)**. Drag the **Ln** to the right, then drag the **Corp Revenue last fiscal yr.** into the **Ln** box.	Perform basic transformations of the Corp Revenue last fiscal yr. attribute.
15	Do the same for the other fields: **Est. Product-Service Spend**, **No of Local employees**, and **No of Years Purchase**.	Continue transforming for the other attributes.
16	Now add a **Page3** and drag another **Cluster** object and add the **Ln** fields you just computed.	Cluster the transformed attributes.
17	The Clustering now should look like Figure 3.3.	Observe the differences between the transformed and untransformed clustered attributes.
18	Save Report1 or save it using another name that you would like to give your report.	

Now let's look at each step in more detail for this example. Make sure that you have all the data sets from this book loaded in your account or in SAS Viya's Public area. These are permanent data locations as they will only be deleted if you select to do so.

Step 1

Loading data into SAS Viya is easy to accomplish. With the CUSTOMERS data set placed in either your account location on the SAS Viya server or in the Public area, log in to SAS Viya with your credentials. In the upper left corner of your browser, you should see the ≡ icon (sometimes called a "hamburger"). Click on that for the Analytics Life Cycle menu and select the menu item called **Manage Data**. You should see some data sets in the **Available** column. Navigate to the **Data Sources** and then find the path to Public or path name that your SAS Administrator has selected for you and/or your team. Click on the data set called **CUSTOMERS**. In the **Details** tab, you should see all the attribute columns listed with their type, raw length, formatted length, and so on. You can sample the data and even run a profile of the data set prior to visualizing, reporting, or analytics if desired. Now, right-click the **CUSTOMERS** data and select the **Load** menu

Figure 3.1: Loading Data in SAS Viya Memory for Exploration and Analytics

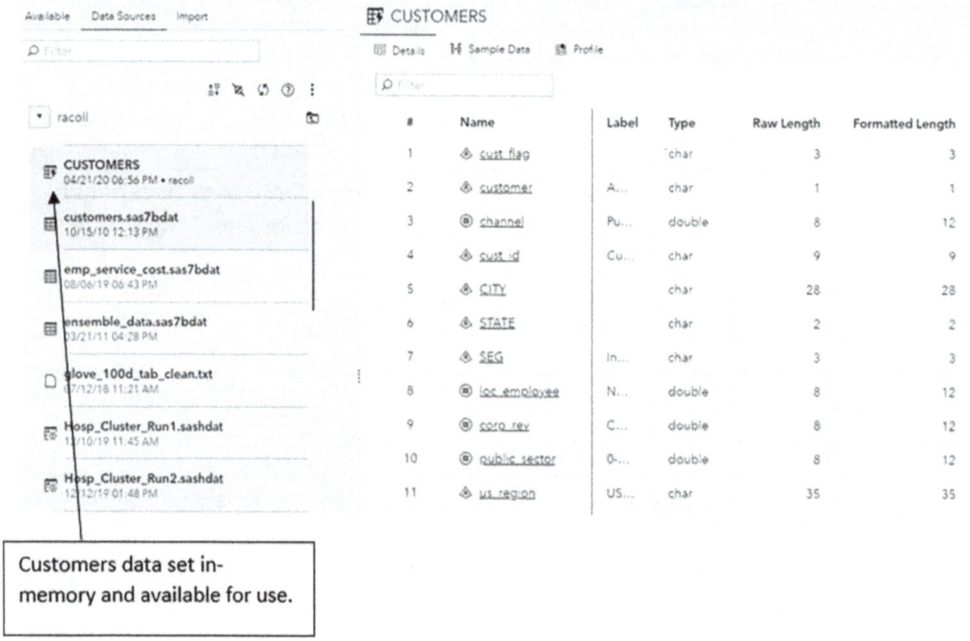

option. After a few seconds, you should see two CUSTOMERS data tables; one with a slightly different icon that indicates it is now in-memory and ready to use. This is shown in Figure 3.1.

Step 2

Now, in the **Analytics Life Cycle** menu ≣ select the **Explore and Visualize** option. Select your folder listed as **My Folder** and then click on the icon that indicates **New Report**.

Step 3

On the far left are four icons: Data, Objects, Suggest, Outline. Click the **Data** icon, and you should see a list of data sets that are in-memory in the Available listing. Click the **CUSTOMERS** data.

Step 4

Now that you have located the CUSTOMERS data set, click on the one with the lightning bolt icon, and you should see the list of fields at the right. Select **Profile** and then run the Profile.

Figure 3.2: Partial View of CUSTOMERS Profile

Column	Unique	Null	Blank	Pattern Count	Mean	Median	Mode	Standa... !
⌖ CITY	6.00% (6,331)			178			NEW...	
⊕ PURCHFST	0.02% (19)				1,992.12	1,998.00	1,998.00	90.16
⊕ PURCHLST	0.02% (19)				1,995.73	2,000.00	2,001.00	90.21
⊕ Prod A	2.40% (254)	89.97% (94...			12.76	2.00	1.00	50.28
⊕ Prod A Opt	3.54% (1,152)	69.17% (...			74.11	7.00	1.00	392.09
⊕ Prod B	2.58% (1,515)	44.34...			79.62	11.00	1.00	413.46
⊕ Prod B Opt	1.46% (1,219)	20.71% (2...			42.27	6.00	1.00	252.47
⊕ Prod C	0.68% (721)	<0.01% (3)			15.35	3.00	1.00	91.75
⊕ Prod D	1.69% (1,508)	15.50% (16,...			58.71	9.00	1.00	250.77
⊕ Prod E	0.98% (119)	88.52% (93,...			4.28	1.00	1.00	29.35
⊕ Prod E Opt	2.09% (336)	84.75% (89...			17.40	4.00	1.00	90.88
⊕ Prod F	1.50% (836)	47.10...			35.79	4.00	1.00	500.56
⊕ Prod G	1.86% (1,118)	43.08...			47.21	5.00	1.00	672.35
⊕ Prod H	6.88% (1,157)	84.06% (88...			313.12	4.00	1.00	2,980.89
⊕ Prod I	0.44% (235)	49.47...			5.00	2.00	1.00	17.85

Step 5

When you run the profile, you should see something like Figure 3.2.

The profile gives some basic statistics (run on a sampling of the data and not the entire data set). You can see how many missing observations exist, the mean, median, and other descriptive statistics. This profile is useful when first becoming acquainted with a data set.

Step 6

After you have run the profile, select OK. The data set is now ready to use, and the data is placed in a default report named **Report1**.

Step 7

If you scroll down the list of data items, you should see a **Measures** listing. Measures are numeric attributes. The default setting for data attributes is to show the label of the attribute if there is a non-empty label for each attribute. Right-click on the attributed listed as: **0- No, 1=Yes**. This is a field for Public Sector. A "1" means that the observation is a public sector organization. Drag the **Public Sector** column onto **Report1** at the right. You should now see a bar chart of the public sector levels 0 and 1 respectively.

Step 8

Now, click the + that is next Page1, and that will add a Page2 to your new report.

Step 9

Click the **Objects** icon that is just under the Data icon. There should be a listing of all types of available objects that you can use to visualize, analyze, and build reports, dashboards, and the like.

Step 10

Scroll down the list of **Object** items until you reach the objects listed under **Visual Statistics** group. Find and select the **Cluster** object and drag it onto **Page2** on the right. We will select a few attributes with which to populate this cluster application.

Step 11

At the far-right hand side of the Cluster object on Page2, you should see another listing of cluster options, roles, action, rules, filters, and ranks. Position your pointer over the blank cluster object and click on it. On the **Data Roles** object, you should see a listing of all numeric variables. Select the following four variables in which to cluster: **Corporate Revenue last fiscal yr.**, **Estimate Product-Service Spend**, **No of local employees**, and **No of Yrs Purchase**.

Step 12

Figure 3.3 should look like your Cluster object now on Page2 of your report. The Cluster object has clustered the data set with these four variable attributes. You can see how each variable is distributed in the clusters. Since these variables are highly skewed in nature, we will transform them and see how that changes the cluster visualization.

The clustering object in SAS Visual Statistics shows each attribute plotted against all the other attributes in the top chart of Figure 3.3. Each box is a pair of attributes plotted with the standard deviation of each attribute and the correlation is calculated. The bottom chart of Figure 3.3 shows each cluster number at the left and how each of the four attributes is distributed in each of the clusters and a color is assigned to each cluster. You can trace how each attribute contributes to each cluster.

Step 13

Now some of the data that we have selected are somewhat skewed distributions. For example, take the upper left quadrant of Figure 3.3 that is corporate revenue for the last year.

Figure 3.3: Visual Statistics: Simple Clustering of Four Attributes (Object on Page2 of Report1 Exploration)

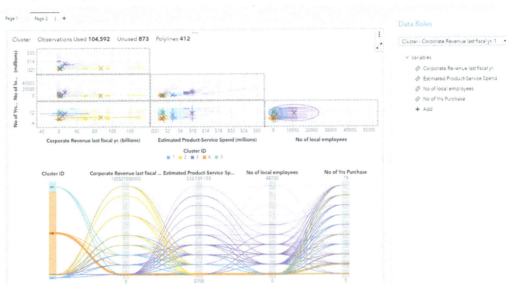

Cluster 2 mostly appears at large bulk above $150B while other parts of cluster 2 are around $110B or less. If we were to plot the distribution of Corporate Revenue for this last year, we'd see a plot something like Figure 3.3a below. I went to Page1 of the report and dragged the Corporate Revenue last year to the right of the Public Sector chart and you obtain two plots side by side on Page1. If you use the expanding icon ⤢ it will expand the chart to the full page. Clicking on it again will restore it so you will see both charts again. If you notice the distribution of Figure 3.3a, the values range from 0 to about $186B. That is quite a range! When data is in this type of distribution, calculating a distance metric is difficult since the data is spread over so large a range. (See Appendix 2 for a detailed description of how clustering calculates distances.) This suggests that we might want to transform this revenue data.

Step 14

So, now click the **Data** icon at the far left and then on the **+ New data item** icon. It should give you a list of options; select **Calculated item**. A new window pops up titled **New Calculated Item**. In the **Name** field in the upper left you can rename the new calculation as **Ln_corp_revenue** for the natural logarithm of the revenue. Now click the **Operators** then select the **Numeric (advanced)** pull-down list. Drag the **Ln operator** onto the right side so that it shows as Ln **(number)**. Then click back on **Data items** and expand the **Numeric** data elements. Select the **Corporate Revenue last fiscal yr.** and drag it to the **number** inside the box of between the ( ). This will take the natural logarithm of this data item and call it **Ln_corp_revenue**. Select the blue **OK** icon to accept this new calculation.

Figure 3.3a: Distribution of Corporate Revenue This Year

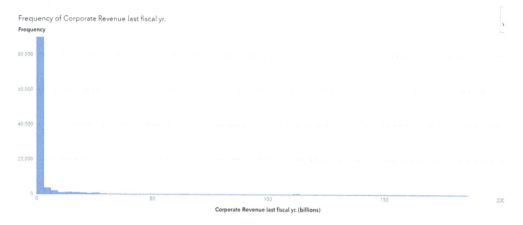

Frequency of Corporate Revenue last fiscal yr.

Step 15

Now do the same for **Est. Product Service Spend, No of Local employees,** and **No of Years Purchase**.

Step 16

Add a Page3 to your report, drag another cluster object onto the blank area, and then add the four newly transformed items in the **Roles** icon at the far right so that you've select each four items as in the **Add Data Items** window below in Figure 3.4.

The clustering should look like Figure 3.5. Depending on the order you created each of the calculated transformations, the total number of observations might be slightly different from what you see in Figure 3.5. The reason for that is that taking the natural logarithm or the base 10 logarithm on these fields will encounter negative numbers in this data set. Since you can't take the log of a negative number, that calculation in a row with a negative value will be left

Figure 3.4: Selection of Four Transformed Attributes for Clustering

˅ Variables

　　　◈ Ln_corp_revenue

　　　◈ Ln_Est_Spend

　　　◈ Ln_Local_Employee

　　　◈ Ln_Num_Yrs_Purcha...

　　＋ Add

Figure 3.5: Visual Statistics: Clustering of Transform Attributes (Object on Page3 of the Report1)

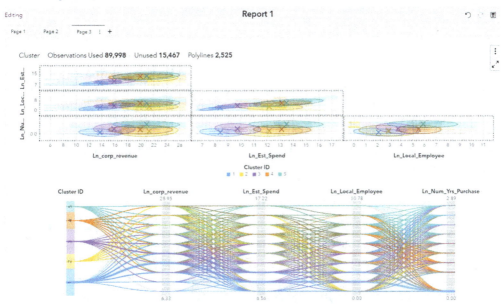

missing! If you want to circumvent that, we can re-edit (do this on your own) each of the four calculations by adding a constant value to the attribute prior to taking the log. In the formula form, we are calculating Ln (Corp Revenue last yr + C) where C is a constant that will shift the data so that all values will be positive ones. The value of C will need to be equal to or greater than the largest negative value of that data item attribute. We will discuss transforms and their impact on clustering in the next chapter.

Step 17

If you observe the new clustering in Figure 3.5 and compare it to the one you did in Figure 3.3, you will notice a much more even spread of the values in the cluster. We will expand on how to profile each cluster more when we do the next task of clustering in SAS Visual Data Mining and Machine Learning in Model Studio in the next task.

Step 18

Save your report by clicking on the ▣ icon in the upper right corner of your report. This will save all your calculated items and any edits or additions that you did in this report.

> **Key Message:** Transforming data attributes will greatly affect how the clustering algorithm measures the distances from one observation to another and to the mean or median cluster value.

Task 2: Using SAS Visual Data Mining and Machine Learning (VDMML) in Model Studio

We now come to Task 2 where we will use Model Studio. Model Studio is a GUI tool within SAS Viya. Its interface is designed for a number of both supervised and unsupervised analytic techniques, and if you combine it with all the SAS procedures and action sets, you have a large arsenal of capabilities not to mention adding open-source languages like R, Python, Lua, and Java. In this exercise, we will use the CUSTOMERS data set as we did in the first task. With Model Studio, you develop data mining and machine learning process flows that SAS calls "pipelines."

Table 3.3 shows the brief outline of steps that we will take to accomplish more advanced capabilities in clustering.

Table 3.3: Clustering Customer Data in Model Studio – Task 2

Step Number	Brief Process Step Description	Brief Rationale
1	Load the CUSTOMERS data set into active memory.	Ensure that the CUSTOMERS data set is available for computations and analysis.
2	Go to the **Analytics Life Cycle** menu and select **Build Models**.	Select model development in the **Build Models** menu that start Model Studio.
3	Create a new project **called Customer Segmentation** and select a target variable **Tot_revenue**. Set the **Channel** variable level to **Ordinal**. Only have the following variables: **cust_id = ID, corp_rev = Input, est_spend = Input, public_sector = Input (level = Binary), Rev_thisyr = Input, and Yrs_purch = Input**.	Set up a Model Studio project, set certain attributes, and reject others to not be used in this project.
4	Right-click and add a **Transformations** node. In the node options for **Interval Inputs**, select **Log transform**. Everything else is set to default values.	Transform revenue attributes using a natural Log transform function. This will help reduce the very wide spread of the revenue data.
5	Add a **Cluster** node under the Transformations node.	Cluster node added with certain option settings.
6	Add a **Segment Profile** node and keep all the default settings. Run this node.	A Segment Profile node will assist in the cluster profiles generated from the statistics of attributes with each cluster.
7	From the **Cluster** node, add a child node and select **Score Data** node from the **Misc** category.	Score node allows the model application to be run on an additional data set or the full data set.

Step 1

Like in the first task, make sure the CUSTOMERS data set is in active memory so that it will be ready to use in your Model Studio project.

Step 2

Navigate to the **Analytics Life Cycle** menu and select the **Build Models** option. This option can enable you to build in SAS VDMML, Visual Forecasting, Visual Text Analytics, and Credit Scoring if you have licenses for those.

Step 3

At the right, select the icon called **New Project** and enter in the name of your project. I called my project **Customer Segmentation** – I know, not too original – but you can select any name that you like. As you see in Figure 3.6, fill in a name and select the **CUSTOMERS** data set. Mine is in a different library, but yours might be in the Public library, which is typically common to all SAS users on the system.

Figure 3.6: New Project Window

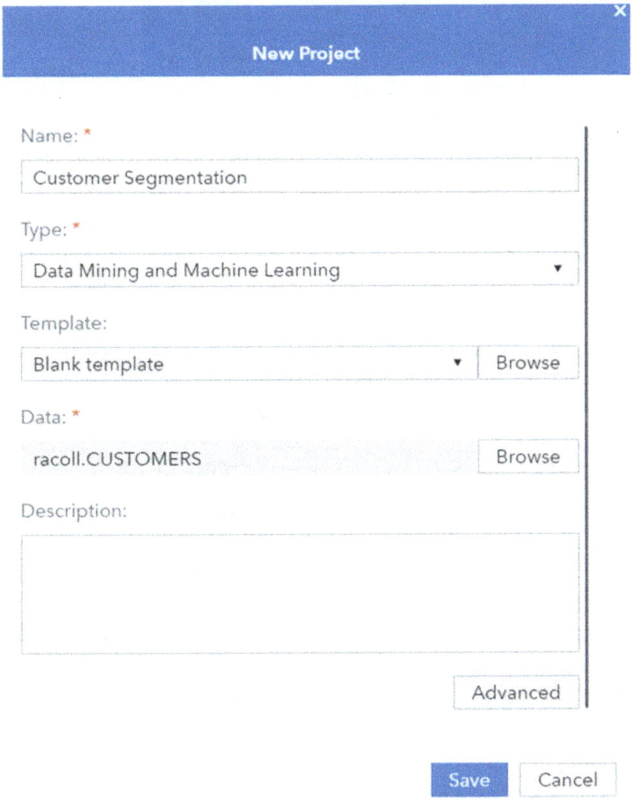

Figure 3.7: Variables Used in This Project

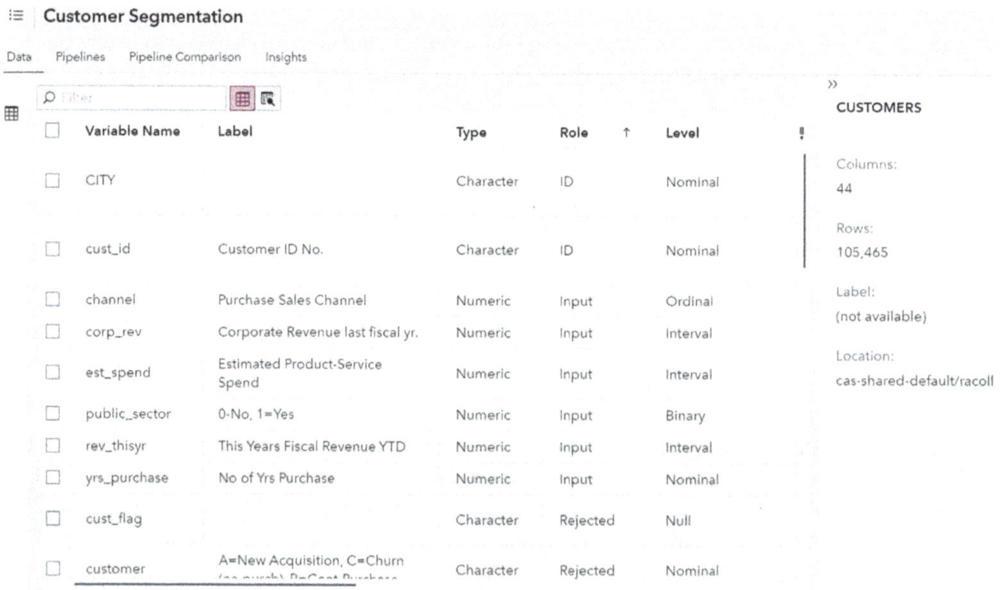

≔   Customer Segmentation

Data   Pipelines   Pipeline Comparison   Insights

	Variable Name	Label	Type	Role ↑	Level	!
☐	CITY		Character	ID	Nominal	
☐	cust_id	Customer ID No.	Character	ID	Nominal	
☐	channel	Purchase Sales Channel	Numeric	Input	Ordinal	
☐	corp_rev	Corporate Revenue last fiscal yr.	Numeric	Input	Interval	
☐	est_spend	Estimated Product-Service Spend	Numeric	Input	Interval	
☐	public_sector	0-No, 1=Yes	Numeric	Input	Binary	
☐	rev_thisyr	This Years Fiscal Revenue YTD	Numeric	Input	Interval	
☐	yrs_purchase	No of Yrs Purchase	Numeric	Input	Nominal	
☐	cust_flag		Character	Rejected	Null	
☐	customer	A=New Acquisition, C=Churn (no purch) R=Cont Purchase	Character	Rejected	Nominal	

CUSTOMERS

Columns:
44

Rows:
105,465

Label:
(not available)

Location:
cas-shared-default/racoll

Keep the **Type** of analytic category set to **Data Mining and Machine Learning** and the template blank as well. There are several default templates to choose from, and new custom templates can be added too. You'll notice that the first tab of the project is the Data tab, and you will need to select a Target variable, even though in this particular case clustering doesn't need any target variables since it is an unsupervised method. Be sure to select only the following variables in this task. Check the box at the left of the variable **tot_revenue**, and at the right side give this variable a role of **Target**. Figure 3.7 shows the remainder of the variables that you will need as input variables, and the remainder should be rejected.

Step 4

Select the **Pipelines** tab and you should now see the Data node at the top of the pipeline area. Now we'll add in a Transformations node to the pipeline diagram. Right-click the **Data** node and select **Add Child Node**, then **Data Mining Preprocessing**, and then **Transformations**. With the Transformations node highlighted, click the **Node options** icon 🗔 located at the right-hand side of the pipeline window. In the **Transformations node options**, pull down the **Interval Inputs** and select the **Log transform**. This will then perform a log transform on all Input variables. It will automatically add 1 to the data variable and then take the log. When the node completes, you should see a green check mark on the node. Right-click the **Transform** node and select **Results**. Figure 3.8 should look like the results that you obtain as well.

Figure 3.8: Transformations Node Results

Summary   Output Data

Input Variable Statistics						Transformed Variables Summary				
Name	Variable ...	Number ...	Percent ...	Minimum		Transformed ...	Method	Input Var...	Formula	Variable L...
channel	ORDINAL	0	0	.		LOG_corp_rev	LOG	corp_rev	log('corp_r ev'n + 1)	INTERVAL
corp_rev	INTERVAL	533	0.8423	0		LOG_est_spend	LOG	est_spend	log('est_sp end'n + 1)	INTERVAL
est_spend	INTERVAL	0	0	815.4503						
public_sect or	BINARY	0	0	.		LOG_rev_thisyr	LOG	rev_thisyr	log('rev_thi syr'n + 1783027.88)	INTERVAL
rev_thisyr	INTERVAL	0	0	-1,783,026. 8800						
yrs_purchas e	NOMINAL	0	0	.						

Step 5

We're now ready to add in the Clustering node. Again, right-click the **Transformations** node and select **Add Child Node**, then **Data Mining Preprocessing**, and then **Clustering**. Now in the Clustering node properties sheet, we'll need to make a few changes to the default settings. The topmost setting is the **Cluster Initialization**. This option enables you to select the algorithm for how cluster seeds are initialized. There are currently two selections in k-means clustering initialization, **Random** and **Forgy**. Forgy is a default selection, and we'll leave that selection. The Forgy algorithm selects random observations from the input data set as a selection for initial cluster centroids.

Next, select the **Interval Inputs** drop-down and set the **Standardization** method to **Range**. Range takes the maximum value and subtracts the minimum value of each variable and selects the that range value to form a scale of 0 to 1 based on the range. The other method uses a Z-score to determine the scale from 0 to 1. If you select **None**, then no scaling is performed and typically that isn't a good option to choose. There are few data sets that have their data already standardized, and in that situation the "None" selection is desired since it was done outside of the Model Studio environment. The **Similarity** distance setting is left at the default setting of **Euclidean distance**. In the **Class Inputs** drop-down, the missing class inputs is left at the **Exclude** setting. That means that any observation that has a missing value for any input variables will drop the entire observation for the analysis. The **Similarity** distance for Class Inputs is left at the default setting of **Binary**.

Now we come to the **Number of Clusters Estimation** setting. We are going to select five clusters as the **User Specify** setting. In the next chapter, we will review why we are setting the number of clusters to 5 by using a visualization technique to observe the potential data structure that might suggest the number of clusters. The **Stop Criterion** is left at default setting and the maximum number of iterations is 10. In the **Scored Output Roles**, we'll want the **Cluster Variable** role to be set to **Segment** and the **Cluster Distance** role as **Input**. That allows the capability of analyzing the cluster distances for each observation. Now you can run the Cluster node. When you open the Cluster node results, your output should look very similar to Figure 3.9.

Figure 3.9: Cluster Node Partial Output Results

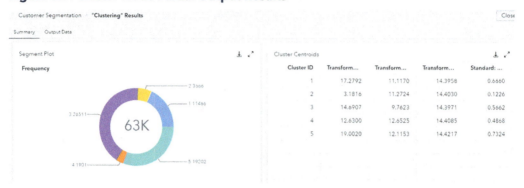

Notice that the number of observations used is about 63,000 since the default setting for the Training data is a random selection of 60% of the total observations.

Step 6

We will now apply this trained Cluster model onto the entire data of 105,000 observations. Right-click the **Cluster** node and select **Add child node**, then **Miscellaneous**, and then **Score Data**. Set the Score Data node options as in Figure 3.10 below apart from the Output library to one you select such as Public or your own assigned library.

Figure 3.10: Score Node Options Settings

Score Data   ○ ▣ ⑦

Description:

Scores a table using the score code
generated by predecessor nodes and
saves the scored table to a CAS library.

⌄ Score Data

   Table name:

    racoll.CUSTOMERS   [Browse]

⌄ Output Data

   Output library:

    racoll   [Browse]

   Table name:

    Score_Cust_Clusters

   ☑ Save table

   ☑ Replace existing table

   ☑ Promote table

   ☐ Drop rejected variables

This node will apply the score code generated to the 105,000 observations and output to the desired table name and library.

Step 7

To complete our Pipeline flow, again right-click the **Clustering** node and select **Add child node**, **Miscellaneous**, and then **Segment Profile**. You can leave all the Segment Profile node options at their default settings. When you open the Segment Profile results, you should see something similar to Figure 3.11.

You can now select various input variables in three of the windows to observe each variable's contribution to each cluster. Figure 3.12 shows the complete Pipeline diagram.

You might want to try adding a new Pipeline to the Cluster Segmentation project and change some of the transformations, other input variables, and the like to see how that affects and changes the cluster solution.

Segmentation Guidelines and Insights

In segmentation, particularly in the marketing and CRM business contexts, there typically isn't a single correct answer of how many segments there are in a data set. What usually happens in marketing is answering the question "Do we have the right number of offers, messaging, and

Figure 3.11: Segment Profile Node Results Window

Figure 3.12: Completed Cluster Pipeline Diagram

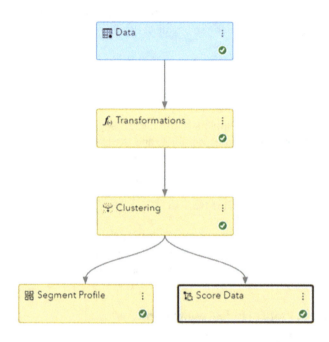

creatives for the segments at hand?" Take, for example, the company Amazon. Amazon has so many items to offer its customers; however, if you take a different industry like Telecom, the offers, messaging, and creatives are seriously limited by comparison. If a Telecom company wishes to give specific offers and messages to 20 million customers, then it would need 20 million very specific offer and message combinations that meet its customers' needs or desires. Since the number of item-offers-message combinations doesn't reach 20 million, there is probably a number of segmentation scenarios that would. If the Telecom company places its customers in 12 behavioral and demographic segments, then it would need 12 such combinations, which is a much more likely candidate. However, if the company offered 150 segments, could marketing come up with 150 very specific and unique offers? Perhaps not. In this context, the number of segments that will drive the needs in marketing is what marketing has to offer and how many messaging and creatives it can practically develop. Still, it might be good to know just how many "true" segments exist with a specific set of variable attributes. We will explore that a bit more in the next chapter by discussing the visualization of underlying segment structures in your data.

Task 3: Using Programming in SAS Studio with CAS and Procedures

We now come to the last of the three tasks in this chapter, and we will use SAS code to develop a cluster solution. This offers the maximum flexibility in analytic capability. Since not every possible option is available in each of the Model Studio nodes, we can access any and all of the potential options in the SAS procedures and action sets along with DATA step and CAS utility coding. We can even use SAS macros to help automate any repetitive data preparation steps that we might desire.

There are several clustering methods available for programming at present in SAS Viya. These are:

- PROC MBC in SAS Visual Statistics, which is the front end to the mbcFit action set
- PROC GMM in SAS VDMML, which is the front end to the gmm action set
- The kClus action set that perform k-means clustering

We will be using the first two procedures because the Cluster node in Model Studio uses the kClus action set. In SAS 9.4, the SAS/STAT module also contains the following five clustering procedures as well: CLUSTER, FASTCLUS, MODECLUS, ACECLUS, and FASTKNN. These SAS/STAT procedures can also be used in SAS Viya; however, the computing will not be in-memory. Also, other methods will support segmentation such as in SAS VDMML using any of the decision tree techniques that can also be used as a segmentation rather than a predictive model (Collica 2017). We will visit that capability when we discuss ensemble segmentation methods in Chapter 6.

The MBC procedure that uses the mbcFit action set is a multivariate Gaussian mixture modeling technique that can use both unsupervised and semi-supervised clustering of data sets. The MBC procedure does assume that the data are continuous, and it does not support categorical or class variables. However, categorical or class variables can be made to be numeric via binning or custom transforms. The traditional goal of clustering and segmentation has been to find groups of observations that are similar by some measurement, often using a distance metric (Collica 2017). Methods that use distance-based metrics appeal to business users because they are rather intuitive by nature, but the methods cannot answer fundamental questions about the number of clusters, the suitability of certain cluster structures, or the method of dealing with outlier observations. In contrast, model-based clustering (MBC) models the set of observations by using a mixture of specific distributions. Using a model in this context, the quality of these clusters and cluster memberships can be estimated by parameters in formal statistical methods. The MBC procedure and action set implements these model-based clustering methods using a mixture of multivariate Gaussian distributions. The MBC procedure allows for a noise or disturbance component and automatic model selection through the use of information criteria. It also provides for output scoring of new input data sets as well. In addition, PROC MBC uses the EM algorithm (SAS Institute 2019) to find parameter estimates whereas other SAS procedures such as FMM and HPFMM (Finite Mixture Modeling) use traditional maximum likelihood and Bayesian techniques.

So, let's get started with our third task, and we will start with the MBC procedure. We will use SAS Studio as our programming interface. Table 3.4 shows the set of steps and the brief rationale for each step that we will take.

Step 1

After you have loaded the CUSTOMERS data and it is in-memory and ready for use, go to the **Analytics Life Cycle** main menu and select the "**Develop SAS Code** option. That should take you to a screen that looks like the one in Figure 3.13.

Table 3.4: Using SAS Programming in SAS Studio for Clustering and Segmentation

Step Number	Brief Process Step Description	Brief Rationale
1	Load the CUSTOMERS data set into active memory.	Make data available for computations and analysis.
2	Start a new SAS Program tab.	Start new program window for SAS statements to be submitted to the CAS server.
3	Statements to start a connect to the CAS server.	Start a CAS in-memory session.
4	Data sampling and DATA step for preparing sample for the MBC analysis.	Sample the CUSTOMERS data set and set up transformations of variables.
5	Analyze sample data for scaling of revenue data.	Transform and scale the revenue attributes as they span several orders of magnitude (power of 10) in ranges.
6	Scale and transform each of the revenues according to the analysis in step 5.	SAS DATA step to transform and scale revenue attributes so that they are as close to a normal distribution as possible.
7	Use the MBC procedure to perform clusters on the sample data set.	Use a multi-variate clustering procedure to assess the likelihood of cluster membership.
8	Use the MBC score action set to apply scores on the CUSTOMERS data set.	Score the MBC clustering on the entire CUSTOMERS data set.
9	Re-sample the final MBC clusters for visualization using the t-SNE algorithm.	Prepare the sampled final data set scores for visualization.
10	Final t-SNE code and graph output.	Visualization of cluster samples.

Figure 3.13: SAS Studio Start Page

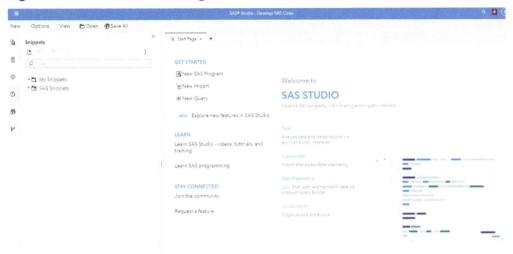

Step 2

Click on the **+** right next the Start Page. This opens a new program window called **Program.sas**. The default window layout for the Program Editor window is a vertical arrangement. You can select the arrangement that you like by clicking on the three vertical dots ⋮ at the right-hand side of the program window, then select **Tab** layout and select either **Single**, **Vertical**, or **Horizontal** splits. Throughout this book, I use the Single layout, but you can select the one you like the best.

Step 3

Write the following statements in the Program Editor, which will start your in-memory session with the cloud analytic server (CAS).

```
cas mySession sessopts=(caslib=casuser timeout=2400 locale="en_US"
metrics='true');

caslib _all_ assign;
```

The CAS statement starts the session called "mySession", and you can name your session as desired. The default CAS library where we will store working data sets is called CASUSER; however, you can call that library what you like, and this is similar to a SAS 9.4 LIBNAME statement. If your editor remains idle it, the time-out is the number of seconds after which your CAS session will end and you will need to resubmit the CAS statements again to re-activate it. Know that if that happens, anything in your CASUSER library will be lost. The "_all_ assign" in the

CASLIB statement indicates that all available CAS SAS libraries will be assigned such as the PUBLIC library and any other libraries that your SAS System Administrator has set up and configured in SAS Viya.

Step 4

```
/* Stratify the CUSTOMERS data set for sampling */
ods noproctitle;
proc partition data=RACOLL.CUSTOMERS samppct=30 seed=13712;
      by channel public_sector us_region RFM;
      output out=CASUSER.cust_samp copyvars=(cust_id loc_employee channel
corp_rev tot_revenue
            rev_thisyr us_region public_sector yrs_purchase est_spend
RFM tot_revenue);
run;
```

We will now use the PARTITION procedure (you could also use the "stratify" action set as the procedure that runs that action) to randomly sample and stratify at a percentage of about 30% of the 105,000 customers. I used a seed value in the statement that helps ensure that we get the same results each time. The stratification is being done by Channel, Public Sector, US region, and the RFM segments. The output statement saves into my CASUSER library. You should use the library for your data source such as PUBLIC instead of my library RACOLL. If your CAS working library is something other than CASUER, then replace that in the output statement.

Step 5

```
ods noproctitle;
/*** Analyze numeric variables ***/
title "Descriptive Statistics for Numeric Variables";
proc means data=CASUSER.CUST_SAMP n nmiss min mean median max std;
      var est_spend corp_rev tot_revenue rev_thisyr yrs_purchase;
run;
title 'Distribution Analysis';
proc univariate data=CASUSER.CUST_SAMP noprint;
      histogram est_spend corp_rev tot_revenue rev_thisyr yrs_purchase;
run; title;
```

We need to transform and scale the revenue data at this step. When you run this code, you should see in the histogram plots that the distribution of the revenues is very skewed. Even though clustering isn't considered a traditional statistical technique, the general "location" of the bulk of data will be needed as almost all clustering techniques are distance-based. Measuring distance in these revenue attributes requires that the mean and median be relatively close together, and when not transformed, they are not even close to each other. Also, some of the data has negative revenues because these were returns, so we will need to scale them to try to push most of the distribution to positive numbers because we cannot take the log of a negative number.

Step 6

```
/* transform revenue data */
data casuser.cust_samp;
set casuser.cust_samp;
    log_est_spend = log(est_spend + 1);
    log_rev_thisyr = log(rev_thisyr + 1227159);
    log_corp_rev = log(corp_rev + 10);
    log_tot_rev = log(tot_revenue + 2311623);
run;
```

In the DATA step in the code above, we will transform and scale these revenues. The value of $1,227,159 is the rounded minimum (negative) of the REV_THISYR attribute, and $2,311,623 is the rounded minimum of the TOT_REVENUE attribute. The minimums of the other variables are 0, so we can just add a small amount so that the log(0) won't be a value of 1.

Step 7

At this point, we come to the stage where we will perform the MBC clustering. We will store the model in a CAS data set called "mbc_score".

```
/* Proc MBC Analysis */
title 'MBC Cluster Analysis and Scoring';
proc mbc data=casuser.cust_samp init=kmeans
    covstruct=(all) nclusters=5 seed=13712 ;
    var log_est_spend log_rev_thisyr log_corp_rev log_tot_rev channel
public_sector;
```

```
 output out=casuser.mbc_cluster_out copyvars=(cust_id cust_id loc_employee
channel
          corp_rev tot_revenue rev_thisyr us_region public_sector yrs_
purchase est_spend
          RFM tot_revenue log_est_spend log_rev_thisyr log_corp_rev log_
tot_rev) maxpost;

 store casuser.mbc_score;

run; title;


title 'Distribution of Cluster Membership';

proc freq data=casuser.mbc_cluster_out;

  tables maxpost; run; title;
```

The MAXPOST statement is the final estimated clusters from the mixing probabilities. Notice that the initial cluster seeds are using the k-means algorithm in the MBC statement. We use the seed value so that we can re-create the same results each time we run the analysis. The PROC FREQ code gives us the estimated distribution of the clusters in the sampling data set. Figure 3.14 shows the partial output of the MBC cluster analysis and saving to a scoring data set.

> **Key Message:** The MBC clustering provides the probability of cluster membership whereas in distance-based clustering, the distance from each observation to the cluster center or mid-point is used to define cluster membership.

Figure 3.14: MBC Cluster Analysis and Scoring Partial Output

MBC Cluster Analysis and Scoring

Model Information	
Number of Gaussian Clusters	5
Covariance Structure	EEE
Noise Cluster Present	No
Expectation Technique	EM
Model Selection Criterion	BIC
Initialization Method	K-means
Convergence Test	Log Likelihood
EM Convergence Criterion	1e-05
Singularity Criterion	1e-08
Parameter Criterion	1e-08
Random Seed	13712

Number of Observations Read	31657
Number of Observations Used	31392

(Continued)

Number of Observations Read	31657
Number of Observations Used	31392

Convergence Status

Covariance Structure	Number of Clusters	Number of Factors	Noise Component	Convergence Status
EEE	5	.	N	Convergence criterion (0.00001) met.
CCU	5	1	N	Convergence criterion (0.00001) met.
EEI	5	.	N	Convergence criterion (0.00001) met.
UUC	5	1	N	Convergence criterion (0.00001) met.
CUC	5	1	N	Convergence criterion (0.00001) met.
UCC	5	1	N	Convergence criterion (0.00001) met.
CCC	5	1	N	Convergence criterion (0.00001) met.
VII	5	.	N	Convergence criterion (0.00001) met.
EII	5	.	N	Convergence criterion (0.00001) met.
VVI	5	.	N	At least one covariance matrix is singular. N
UUU	5	1	N	At least one covariance matrix is singular. N
EVV	5	.	N	At least one covariance matrix is singular. N
CUU	5	1	N	At least one covariance matrix is singular. N
EVI	5	.	N	At least one covariance matrix is singular. N
UCU	5	1	N	At least one covariance matrix is singular. N
EEV	5	.	N	At least one covariance matrix is singular. N
VVV	5	.	N	At least one covariance matrix is singular. N

Mixing Probability Estimates for Selected Model

Mixing Component	Mixing Probability
1	0.10625
2	0.18551
3	0.29734
4	0.23602
5	0.17488

Model Selection Summary

Covariance Structure	Number of Clusters	Number of Factors	Noise Component	Number of Parameters	-2 Log L	AIC	AICC	BIC
EEE	5	.	N	55	323711	323821	323822	324281
CCU	5	1	N	46	325707	325799	325799	326183
EEI	5	.	N	40	331304	331384	331384	331718
UUC	5	1	N	69	366448	366586	366586	367162
CUC	5	1	N	45	383135	383225	383226	383601
UCC	5	1	N	65	402567	402697	402697	403240
CCC	5	1	N	41	475577	475659	475659	476001
VII	5	.	N	39	514887	514965	514965	515291
EII	5	.	N	35	524944	525014	525015	525307
VVI	5	.	N	64	.	.	.	.

Let's review the output so that you can understand the basics of the analysis. The covariance structure is labeled with letter codes to indicate the type of covariance matrix being used. Of key importance are the covariance matrices that didn't converge and are considered *singular.* Essentially, each structure code name depicts a different volume, shape, and orientation of the cluster distributions. Only the matrices that converged were used in the final analysis. The mixing probability table gives the probability of mixing for each cluster segment. The last part of the output gives the final fit statistics using log likelihood, AIC/AICC, and BIC criterion for each covariance matrix. The SAS documentation on the MBC procedure gives a good summary of the covariance structures, Gaussian Mixtures, and also contains other references for further study (SAS Institute 2019).

Step 8

Now that we have the model developed on the sample data set, we will use the scoring to develop the clusters on the entire data set. For that we need to use the mbcScore action set. First, we need the final data set that has the transforms and scales like we did on the sample, then the PROC CAS statement that deploys the mbcScore action set.

```
* Apply MBC Cluster Solution to entire Customers data set */

data casuser.all_customers;

  set racoll.customers;

 log_est_spend = log(est_spend + 1);

    log_rev_thisyr = log(rev_thisyr + 1227159);

    log_corp_rev = log(corp_rev + 10);

    log_tot_rev = log(tot_revenue + 2311623);

keep cust_id cust_id loc_employee channel corp_rev tot_revenue rev_thisyr
us_region

      public_sector yrs_purchase est_spend RFM tot_revenue log_est_spend
log_rev_thisyr

log_corp_rev log_tot_rev;

run;


title 'Final MBC Cluster Analysis';

proc cas;

    action mbc.mbcScore /

      table={name='all_customers'}

      restore={name='mbc_score'}

      casOut={name='mbc_cluster_final',replace=true}
```

```
        copyvars={'cust_id','channel','public_sector','log_est_spend',
                'log_rev_thisyr','log_corp_rev','log_tot_rev'}
        maxpost='group'
        nextclus='Cluster_Weight';
    run;
quit; title;


title 'Distribution of Final Cluster Membership';
proc freq data=casuser.mbc_cluster_final;
    tables group; run; title;
```

Step 9

```
/* Visualize clusters using t-SNE */
/* obtain sample between 3 and 5k observations */
proc partition data=casuser.mbc_cluster_final samppct=3.5 seed=13712;
        output out=CASUSER.mbc_cluster_sample copyvars=(cust_id group log_
est_spend log_rev_thisyr log_corp_rev
            log_tot_rev channel public_sector);
run;
```

We use PROC PARTITION again to generate a 3.5% sample of the final scoring data set. This is because the t-SNE algorithm can only handle about 3,000–5,000 observations for visualization – any more observations and in the graphic output it will be difficult to see individual data points, and also the algorithm's convergence will require too much memory. We'll review more on this in the next chapter on visualizations.

Step 10

```
title 'Visualization of Cluster Segments with t-SNE Algorithm';
proc tsne data=casuser.mbc_cluster_sample;
    input log_est_spend log_rev_thisyr log_corp_rev log_tot_rev channel
public_sector;
output out=casuser.mbc_cluster_tsne copyvar=group;
run; title;
```

```
title 'MBC Cluster Membership visualized using t-SNE Algorithm';

proc sgplot data=casuser.mbc_cluster_tsne;

  scatter x=_dim_1_ y=_dim_2_ / group=group markerattrs=(symbol=circle
size=5);

run; title;
```

Figure 3.15 shows the graphic visualization of the t-SNE results. These cluster shapes should be viewed as the relevant underlying structure of these attributes as well as MBC cluster membership.

> **Key Point**: Clustering and t-SNE are performing very different algorithms on the data; the former measures distance from each observation to another and the latter measures the local proximity of each observation for dimension reduction visualization.

Notice how the MBC clustering assigns cluster membership versus how we visualize the clusters using the t-SNE algorithm. The four sub-clusters at the bottom of Figure 3.15 are all part of the Channel attribute. The channel has four levels: 0, 1, 2, and 3. If the MBC algorithm depicted

Figure 3.15: MBC Cluster Visualization in Two Dimensions

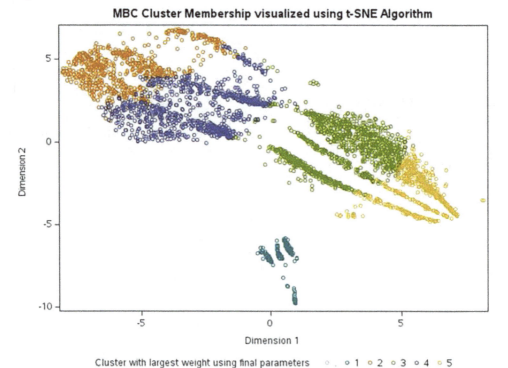

what we visualize, the elongated groups would likely be separate clusters rather than cluster memberships cutting across these striation patterns. We will discuss visualization versus algorithmic cluster groupings in greater detail in the next chapter.

References

Collica, R. S. 2019. *Customer Segmentation and Clustering Using SAS® Enterprise Miner*. Cary, NC: SAS Institute Inc., Chapters 3,4, and 8.

SAS Institute Inc. 2019. *SAS® Visual Statistics 8.5: Procedures*. Cary, NC: SAS Institute Inc. https://go.documentation.sas.com/?cdcId=pgmsascdc&cdcVersion=9.4_3.5&docsetId= casstat&docsetTarget=casstat_mbc_overview.htm&locale=en

Chapter 4: Envisioning Underlying Structure and Clusters in Your Data

What Is t-SNE and How Does It Work?

In this chapter, the ideas and concepts behind visualizing multidimensional data will be briefly reviewed and discussed. The visualization of data, especially when there are many variable attributes and thus multiple dimensions, is an important and sometimes a difficult problem in many domains and industry applications. For example, take the word embedding from natural language processing (NLP) when word-term-phrases and document counts are placed in a matrix. As another example, the number of variable attributes in a consumer demographic database can easily comprise hundreds of columns. Gene expression data has thousands of attribute variables that can possibly lead to hundreds of thousands of columns. The algorithms that process such large and sparse matrices such as singular value decomposition (SVD) or principle components (PC) output data that consists of transformed multidimensional data. Most multidimensional techniques provide methods to display more than two data dimensions; however, the spatial representation of the original untransformed data is typically lost (Van der Maaten and Hinton 2008). Visualizing the transformed dimensions then becomes an arduous task of determining the spatial representation in the transformed data space, which is not anything like the actual untransformed data space. One of the objectives of dimension reduction is to keep as much of the underlying structure of the high-dimensional data as possible in the low-dimensional data space (Van der Maaten and Hinton 2008). The t-distributed stochastic neighbor embedding (t-SNE) algorithm has some very nice properties that PC and SVD don't have – the ability to capture much of the local structure of the high-dimensional data well and reveal the larger, more global groupings (SAS Institute 2019). Some of the key points on t-SNE are as follows:

- t-SNE is a non-linear projection. It uses local relationships (points near each other) to create a map between the high- and low-dimensional space. This allows the algorithm to capture the non-linear data structure.

- The algorithm creates a probability distribution that defines the relationship between points in the high-dimensional space.
- The Student's t-distribution is used to re-create the probability distribution in the low-dimension space that is used for visualizing. This prevents the "crowding" problem where points are too close together in the low dimension.
- The t-SNE algorithm optimizes the embeddings using gradient descent. The cost function is non-convex, meaning that there is some risk of getting stuck in local minimum; however, t-SNE has a few tricks to avoid this issue.

The t-SNE algorithm is a non-linear dimension reduction technique. Our task is to see whether we can visually understand the basic structure of the data prior to performing a cluster model and then use our visual insight to guide the cluster solution and see whether the clustering is similar to the original data visualization using t-SNE. PROC TSNE only accepts numeric data, so any categorical data needs to be converted to numeric. Several methods for doing so include the GLM coding method, deviance, or other methods for categorical to numeric transforms (Collica 2017). The "t" in t-SNE means a Student's t-distribution process. A t-distribution with one degree of freedom is used in the distribution of joint probabilities because the non-linearity produces longer tails, so a t-Distribution has a property that approaches an inverse square law for large pairwise distances in the low-dimensional mapping (Van der Maaten and Hinton 2008). In other words, the t-distributed SNE is better able to handle the non-linear transformation between the high and low dimension spaces. The three key tuning parameters are: 1) learning rate, 2) the maximum number of iterations, and 3) the perplexity – a bandwidth parameter that varies with each data point that closes matches the fixed one the user specifies or the default value of 30 (Kurita 2018).

How Does t-SNE Work?

A high-level description of the t-SNE algorithm steps can be outlined as follows (Kurita 2018):

1. A probability distribution is created that governs the relationships between various neighboring points.
2. Re-create a low-dimensional space that follows a probability t-distribution as best as possible to keep the original data structure in the high dimension.
3. Optimize to find the best mapping relationship using some tuning parameters.
4. The original data set is considered higher dimensional space because the number of attributes is greater in that dimension than after the dimension reduction of t-SNE where the graphic visualization is in two or three dimensions and therefore lower attributes and thus lower number of dimensions to visualize.

t-SNE works to solve the "crowding problem" when you transfer from high- to low-dimension space. For example, take this illustration in Figure 4.1.

When you map the one-dimensional space to two-dimensional space, you have the issue of multiple points at the same medium distance from a certain point. The crowding problem stems from the "curse of dimensionality." In the high-dimensional space, the surface of a sphere grows

Figure 4.1: Illustration of High- to Low-Dimension Space

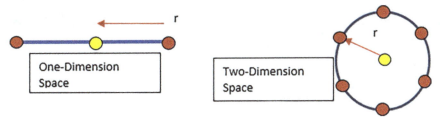

much faster with its radius compared to a sphere in low-dimensional space. Now, let's imagine what happens when these points get mapped into a lower-dimensional space. With a linear or naïve approach, the massive number of medium distance points will try to get pushed in the higher-dimensional space, and therefore points get squashed together and cause the crowding problem (Kurita 2018). The t-SNE algorithm handles this problem by making the optimization routine spread out the medium distance points to prevent crowding.

Another major benefit of t-SNE is stochastic neighbor embedding. The meaning behind "stochastic neighbors" is that there is no clear line between which points are neighbors to other points, and this lack of definitive borders is an advantage because t-SNE can take both global and local structure into account without being restricted by the single-line neighborhood (also known as linear) methods (Kurita 2018).

One of the disadvantages of t-SNE is that the algorithm is non-deterministic. That is, you can run it multiple times and get slightly different results each time. The authors of the t-SNE algorithm have modified the optimization methods to find the local minimum; however, there is no fail-safe method to reach the true global minimum (Kurita 2018). The main issue in the t-SNE analysis is adjusting the perplexity factor such that the visualization is relatively stable. The perplexity is related to the number of neighbors for any one given data point; the typical best practice is to set the value between 5 and 50. In SAS, the TSNE procedure default value of perplexity is set to 30. You might have to experiment a bit and try different perplexity ranges to find the one that is best suited for your application.

> **Key Messages:** t-SNE's main goal is to keep the spatial orientation of the original data set in the reduced dimension data space. The end goal is to be able to visualize the data in the reduced space to observe underlying data structure and potential patterns.

Task 1: Using SAS Studio Programming – Feature Engineering and the Impact on ML Methods

Let's start using our CUSTOMERS data set again. Follow the details that refer to each brief process steps in Table 4.1. In this example, we are using the CUSTOMERS data set again; however, the variables that we chose to transform in Chapter 3 with a simple log transform and linear scaling still didn't quite do the trick for normalizing our data, which is highly skewed. In the fields of data

Table 4.1: Feature Engineering Steps

Step Number	Brief Process Step Description	Brief Rationale
1	Load the CUSTOMERS data set into SAS Viya memory. Copy to your CASUSER session library.	Ensure that the CUSTOMERS data set is available for processing.
2	Sample the CUSTOMERS data set with stratified sampling to ~5–6k observations.	Visualizing more than 5–6k observations isn't very helpful, and the t-SNE will take up too much memory in the mapping process.
3	Feature engineering transforming and scaling of attributes.	Transform and scale the revenue attributes to be visualized.
4	Use the TSNE procedure to visualize the sample data set in two and three dimensions.	
5	General observations of the visualization that t-SNE produced.	

science and machine learning, the analyst who is keen on attribute feature manipulation will often have some of the best models around. This is true for the simple fact that when you feed better and cleaner data to your algorithm of choice, the results will be better in general.

This task gets you started using PROC TSNE for visualizing a data set sample on six variable attributes. Later, we will cluster those same attributes and compare and contrast the t-SNE visualization with the clustering.

Step 1

To load data into memory, you use the **Manage Data** option in the **Analytics Life Cycle** menu at the far left of the browser interface. Right-click the **CUSTOMERS** data set and select **load** to place the data set into memory for use. Now use the SAS DATA step to copy that data into your CASUSER library with the following code. The first three lines open a CAS session in SAS Studio, assign all available CAS libraries, and name your CAS library as CASUSER. Lines 5 to 7 copy the in-memory data to your local CAS CASUSER library session.

```
cas mySession sessopts=(caslib=casuser timeout=2400 locale="en_US"
metrics='true');

caslib _all_ assign;


data casuser.customers;

set racoll.customers;

run;
```

Step 2

Now that the data is available in your CASUSER session library, we need to sample it. We will use the "sampling" CAS action in order to accomplish the sampling.

```
/* Stratify Sample for Visualizations */
proc cas;
   loadactionset "sampling";
   action stratified result= r /table={name="customers",
groupby={"channel","public_sector",
         "us_region","rfm"}}
      samppct=6 samppct2=40 partind="TRUE" seed=3712
    output={casout={name="cust_sample", replace="TRUE"},
copyvars={"cust_id","est_spend","rev_thisyr","corp_rev","tot_
revenue","channel",
      "public_sector"}};
run;
  print r.STRAFreq; run;
quit;

proc freq data=casuser.cust_sample;
  tables _partind_ ; run;
data casuser.cust_sample;
   set casuser.cust_sample;
where _partind_ = 1;
run;
```

The above SAS code calls the "sampling" CAS action, and the data set to sample from is the CUSTOMERS data in the current session. We will stratify on the fields channel, public_sector, us_region, and rfm. The sampling percentage is accomplished by using both the samppct and the samppct2 options. The seed value ensures that we get the same observations if we run the same code multiple times. The print statement prints out the CAS sampling result from the table r.STRAFreq internal table. I also included a PROC FREQ statement to list out the frequency distribution of the _partind_ variable that depicts the sampling levels. The last SAS DATA step code only outputs to CUST_SAMPLE the approximately 6k observations for our analysis.

Step 3

Now comes the feature engineering part where we will need to transform and scale the revenue attributes. The revenues span several powers of 10, so by using the log transform, we will squash that down considerably. We will use a similar scaling to what we did in Chapter 3.

The scaling helps to ensure that we won't take the log of negative values and cause an error and therefore leave that observation blank in the output data set.

```
data casuser.scale_fin_out;

  set casuser.cust_sample;

drop _partind_ ;

    log_est_spend = log(est_spend + 1);

    log_rev_thisyr = log(rev_thisyr + 10);

    log_corp_rev = log(corp_rev + 100);

    log_tot_revenue = log(tot_revenue + 100);

run;
```

Step 4

Now we will use the TSNE procedure to run the algorithm and plot the results. The SAS code below does that and an additional run of t-SNE to see what things look like in three dimensions instead of just two.

```
title 'Random Sample of Customers';

proc tsne data=casuser.scale_fin_out ndimensions=2 seed=13712 ;

    input log_est_spend log_rev_thisyr log_corp_rev log_tot_revenue channel

public_sector;

output out=casuser.tsne_output ;

run;


title 'Random Sample of Customers';

title2 'Plot of Potential Underlying Structure';

proc sgplot data=casuser.tsne_output;

    scatter x=_dim_1_ y=_dim_2_ /markerattrs=(symbol=circle size=5);

run; title; title2;
```

```
title 'Random Sample of Customers 3 Dimensions';

proc tsne data=casuser.scale_fin_out ndimensions=3 seed=13712 ;

   input log_est_spend log_rev_thisyr log_corp_rev log_tot_revenue channel

public_sector;

output out=casuser.tsne_output3;

run;
```

Figure 4.2a shows the two-dimensional plot, and Figure 4.2b is the three-dimensional plot. Figure 4.2b is plotted using SAS JMP (a desktop package). You could also use the SAS procedure called PROC G3D to make a three-dimensional plot as well; however, in JMP, the three-dimensional plot can be rotated interactively. The default settings are used for the tuning in t-SNE; the perplexity default value is 30, the learning rate default is 100, and the maximum iterations is 1000 (Kurita 2018).

Step 5

At this point in our analysis, we should review the charts in Figures 4.2a and Figure 4.2b. In Figure 4.2a, we can easily notice that there are distinctly five main clusters, and one cluster in the lower right corner appears to have two or three sub-clusters. Remember that the t-SNE algorithm

Figure 4.2a: t-SNE of Several Attributes of CUSTOMERS Data Set – Two-Dimensional View

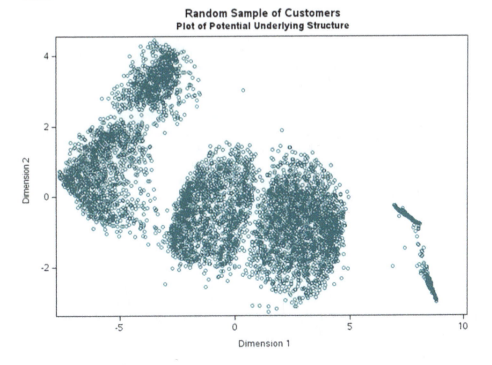

Figure 4.2b: t-SNE of Several Attributes of CUSTOMERS Data Set – Three-Dimensional View

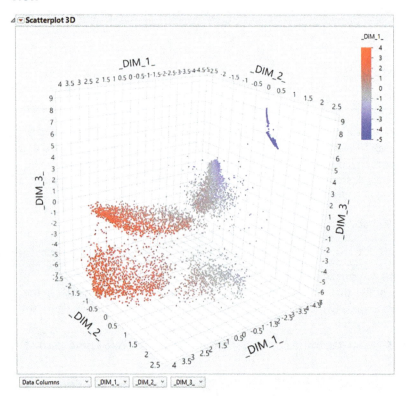

attempts to keep the same spatial relationships of the data points of the lower-dimensional space as in the higher-dimensional space. In Figure 4.2b, the lower-dimensional space has three dimensions, and we do see the third dimension spatially separating the clusters as in dimension one and two. The representation of the dimension reduction of Figure 4.2a compared to the original data that has six dimensions (the six variable attributes selected for the analysis) should be represented in a relatively similar type of plot. In the next task, we will take what we have learned in Task 1 with the six attributes and perform a clustering using 5 as the number of clusters. While the t-SNE algorithm isn't really doing any clustering, it is trying to find other observations for an attribute that is considered a "local" neighborhood. What constitutes a "local" neighborhood depends a bit on some of the tuning parameters of t-SNE.

Task 2: Using SAS Studio Programming – Clustering and t-SNE Comparison

In this next task we will investigate the cluster results from Chapter 3 with this t-SNE to see whether the clustering is giving similar or dissimilar results compared with the dimension reduction visualization.

Table 4.2: Steps for Clustering and t-SNE Visualizations

Step Number	Brief Process Step Description	Brief Rationale
1	From the scored data set we did in Model Studio in Chapter 3, load the scored results data set into SAS Viya memory and copy to your CASUSER library.	Load CUSTOMERS data that has the cluster scoring into SAS Viya memory.
2	Stratify the data set with the clusters and include the _cluster_id_ variable.	Sample the cluster scored data as in Chapter 3. Need to include the _cluster_id_ because this identifies the cluster membership of each observation in the data set.
3	From the stratified sample, re-create the transformed variables in the sample.	Re-create the transformed revenue attribute variables for visualization.
4	Re-run the TSNE procedure on this new sample that has the cluster scores with the same settings as in Chapter 3.	Run the TSNE procedure with the cluster identification in the data set.
5	Plot the results of the t-SNE output but with coloring of the cluster ID	Plot the t-SNE results along with color identification of cluster membership.
6	Compare and contrast the output visualizations from Figure 4.2a and Figure 4.3.	Compare the charts for visual analysis to see how the t-SNE compares with the clustering.
7	Comments and conjecture.	Comments on the analysis.

Step 1

In the Model Studio segmentation pipeline that we did in Chapter 3, the last part scored the entire data set with the five clusters and saved the data set into a permanent CAS library for later use. Now, we will need that data set again, so load that data set into CAS memory.

Step 2

Copying the data set in lines 5 to 7 allows the temporary CAS library CASUSER to hold the data set we will analyze in this task.

```
cas mySession sessopts=(caslib=casuser timeout=2400 locale="en_US"
metrics='true');

caslib _all_ assign;

data casuser.customers;
  set racoll.customers;
run;
/* Stratify Sample for Visualizations */
proc cas;
   loadactionset "sampling";
   action stratified result= r /table={name="customers",
groupby={"channel","public_sector",
          "us_region","rfm"}}
      samppct=6 samppct2=40 partind="TRUE" seed=3712
    output={casout={name="cust_sample", replace="TRUE"},

copyvars={"cust_id","est_spend","rev_thisyr","corp_rev","tot_
revenue","channel",
      "public_sector"}};
run;
  print r.STRAFreq; run;
quit;

proc freq data=casuser.cust_sample;
  tables _partind_ ; run;
```

Step 3

```
data casuser.cust_sample;
   set casuser.cust_sample;
where _partind_ = 1;
run;
data casuser.scale_fin_out;
  set casuser.cust_sample;
```

```
drop _partind_ ;

    log_est_spend = log(est_spend + 1);

    log_rev_thisyr = log(rev_thisyr + 10);

    log_corp_rev = log(corp_rev + 100);

    log_tot_revenue = log(tot_revenue + 100);

run;
```

Step 4

We need to alter the TSNE procedure slightly by including the copyvar= _cluster_id_ statement so that we can compare the clustering versus the t-SNE visualization mapping.

```
title 'Random Sample of Customer Segments';

proc tsne data=casuser.customer_segments ndimensions=2 ;

    input log_est_spend log_rev_thisyr log_corp_rev log_tot_rev channel

public_sector ;

output out=casuser.tsne_cluster_output copyvar = _cluster_id_ ;

run;
```

Step 5

In this next snippet of code, we alter the plot slightly by grouping the cluster ID. The colors are now attributed to the clusters, and the data points are the t-SNE mapping. The second PROC TSNE code performs the same analysis except using three dimensions instead of two.

```
title 'Random Sample of Customers';

title2 'Plot of Potential Underlying Structure';

proc sgplot data=casuser.tsne_output;

    scatter x=_dim_1_ y=_dim_2_ /markerattrs=(symbol=circle size=5);

run; title; title2;


title 'Random Sample of Customers 3 Dimensions';

proc tsne data=casuser.scale_fin_out ndimensions=3 seed=13712 ;

    input log_est_spend log_rev_thisyr log_corp_rev log_tot_revenue channel
```

```
public_sector;

output out=casuser.tsne_output3;

run;
```

Step 6

Let's carefully inspect Figure 4.3 with respect to Figure 4.2a. The first observation we might make is that the overall shapes of the grouped data points are very similar, but not exactly the same as in Figure 4.2a. This is due to the sampling in a distributed computing platform.

When SAS Viya operates with number of computing CPU cores, the in-memory analysis becomes distributed among all the worker nodes of the CPU cores of the compute platform. When this happens sometimes from one session to another even when our code hasn't changed (or seed levels for sampling), the exact observations aren't always preserved identically but are very similar. The bottom of Figure 4.3 shows the three levels (sub-clusters), whereas in Figure 4.2a, we only appear to see two. In the actual data set, this cluster and sub-clusters represent the

Figure 4.3: t-SNE of Several Attributes of Customers Data Set with Coloring ID of Cluster Scores

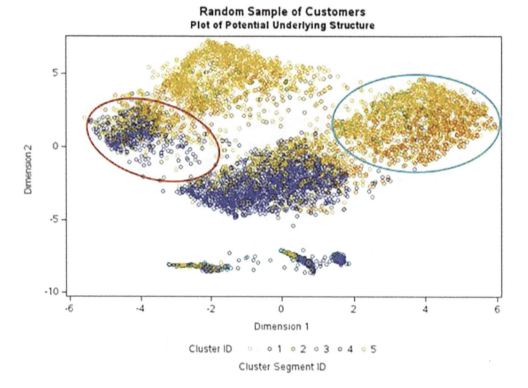

CHANNEL variable that has three levels – 0, 1, and 2. Also, we see that the coloring of points is mixed somewhat in the clusters, so there are some differences in crossing over of cluster membership to the larger structure we observe. The dark red oval in Figure 4.3 shows the blue and orange mixed in this group, whereas the light blue oval on the right shows mixing of orange, yellow, and a few green as well.

While this isn't an exact science, we do see a lot of basic similarities of the cluster membership and the overall t-SNE visualization, so in general we could say that the five major clusters are in general agreement with each other. The t-SNE algorithm is one technique that you might want to use as an aid to pre- and post-clustering or profiling of data sets.

Key Messages: t-SNE is a valuable visualization technique that attempts to allow the underlying spatial data structure in the reduced dimension data space. This capability is an excellent aid for clustering and segmentation in use before and after clustering.

References

Collica, R. S. 2017. *Customer Segmentation and Clustering Using SAS® Enterprise Miner™*. Cary, NC: SAS Institute, Chapters 3, 5, and 6.

SAS Institute, Inc. 2019. "The TSNE Procedure," *SAS® Visual Data Mining and Machine Learning 8.5 Procedures*.
Cary, NC: SAS Institute Inc. https://go.documentation.sas.com/?cdcId=pgmsascdc&cdcVersion=9.4_3.5&docsetId=casml&docsetTarget=casml_tsne_toc.htm&locale=en

Van der Maaten, Laurens, and Geoffrey Hinton. 2008. "Visualizing Data Using t-SNE," *Journal of Machine Learning Research* 9:2579-2605.

Chapter 5: Ensemble Segmentation: Combining Segmentations Algorithmically

Methods of Ensemble Segmentations

The word *ensemble* means to combine, collect, or collaborate. Ensemble models have been in use for quite some time. Typical methods for combining different models of the same target response variable have been reported and classified as bagging or boosting. *Bagging* is short for bootstrap aggregation, and one of the first reported bagging algorithms was by Breiman (1996). The function that combines the models could be to average the results together from different models, find the model with the maximum probability, or to vote for the maximum probability (Berk 2004). In *boosting*, the algorithm attempts to "learn" how to classify a target response variable by "boosting" the weak classifiers to make a stronger classification model once combined. The basic idea of the boosting algorithm is to construct a filtering mechanism so that the majority voting of different estimates combines to settle on a single estimate. A common boosting algorithm is called XGBoost (Chen and Guestrin 2016).

So far, the ensemble methods are a combination of predictive numeric or categorical/nominal models. We now turn our attention to ensemble segmentations and ensemble clustering. In ensemble clusters, the goal is the combine cluster labels, which are symbolic; therefore, one must also solve a *correspondence problem* (Strehl and Ghosh 2002). This *correspondence problem* occurs when two or more segmentations or clusters are combined to form a new segmentation. The goal of this new segmentation is to find the best method to combine them so that the final segmentation has better quality and/or features not found in the original uncombined segmentations (Ghaemi et al. 2009). In essence, the final combined segmentation has the best or most desirable features in the combination than in the input segmentations.

Strehl and Ghosh (2002) used a couple of methods to combine the results of multiple cluster solutions. I find these methods to be esoteric in nature, and the methods do not scale when data becomes large. More recently, I have devised a software patent for combining two more segmentations, and we will use this method described briefly below (Collica 2015).

Ensemble Segmentation

In Figure 5.1, the table describes three different segmentations, each with three levels for simplicity each one is labeled as $\lambda_1 - \lambda_3$. We would like to have a combined segmentation, λ_4, that most effectively combines segmentations 1–3. The outcome response is λ_4 and the input segmentations are λ_{1-3}. Let r be the number of original cluster solutions; in this case, 3. And let λ_n represent the input cluster segmentations and $r=3$. Each x_r represents a single observation where the cluster or segmentation solution will be applied on each of the n data records. One criterion for combining a set of cluster or segmentation solutions would be to maximize the mutual information gain. In information theory, mutual information quantifies the statistical information between shared distributions (Strehl and Ghosh 2002).

Figure 5.1: Combining Segmentations 1-3

λ_4		$\lambda_{(1)}$	$\lambda_{(2)}$	$\lambda_{(3)}$
	x_1	1	3	2
	x_2	1	2	1
	x_3	3	1	1
	x_3	2	3	3
	x_4	?	1	2
	.	.	.	.
	.	.	.	.
	.	.	.	.
	.	.	.	.
	x_n	2	1	2

Another method might be to weight one of the cluster input solutions higher or lower than the others that corresponds to some business objectives. For example, λ_1 might be given a higher weight than λ_2 by multiplying. A weight of 1.5 could be applied to λ_1 whereas λ_2 would have a weight of 1. There could be an almost infinite set of methods and weighting schemes for combining the clusters and segments into a final ensemble solution. Perhaps a more optimal set of combinations can be addressed with the proper business objectives and goals for the use of the final cluster or segmentation solution. For a formal argument on the effectiveness of cluster ensembles, refer to Topchy et al. (2004).

In the patent "System and Method for Combining Segmentation Data," I have a two-step process for combining and evaluating the segmentation. The first step is the combination step, which can comprise several potential algorithms. The second step is for evaluation and adjustment, which uses Bayesian methods. Potential algorithms for combination include clustering, decision trees, self-organizing neural networks, and others. If you recall from Bayes' theorem, Equation 5.1 is the Naïve Bayes formula (Mitchell 1997).

$$P(\lambda_i \mid x) = \frac{P(x \mid \lambda_i)P(\lambda_i)}{P(x)} \qquad\qquad 5.1$$

This formula basically states that the Naïve Bayes classifier is based on the simple assumption that the attribute values are independent given the target value. The application in ensemble segmentation is that each of the input segmentations to be combined need to independent from each other and therefore come from different sources of data. As said earlier, one way to determine the best possible combination is to hypothesize that best refers to the most probable outcome (Mitchell 1997). Bayesian methods are known for their ability to determine the most probable outcome based on input data from varied sources of information; in our case, different segmentations. Probability estimation using Naïve Bayes' theorem is described nicely by Domingos and Lowd (2005). Figure 5.2 shows a general flow the process depicting two input segmentations using k-means clustering followed by Naïve Bayes estimation (Collica 2015).

Basically, what this process method entails is to first combine two or more segmentations using an algorithm such as k-means clustering and then evaluating the newly created segmentation using a Bayesian estimation technique followed by any probability level adjustments. After this process, the newly defined segmentation will be profiled to best understand what the final outcome segmentation has learned from the input segmentations. The flow diagram in Figure 5.2 depicts estimating the probability using a Naïve Bayesian approach; however, the approach can be somewhat more general. A Bayesian network estimation of the newly formed clusters could be a general Bayesian network, naïve, or a parent-child structure, and so on (SAS Institute, "B-Net Procedure" 2019).

Figure 5.2: Flow Diagram of Ensemble Segmentation with Naïve Bayes

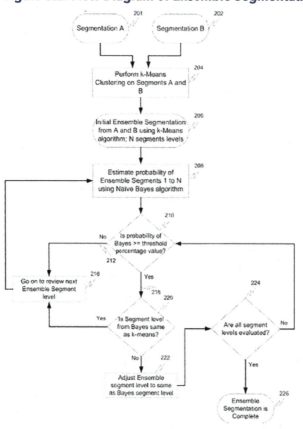

Source: Collica, R. S., "System and Method for Combining Segmentation Data," US Patent, Applicant: SAS Institute Inc., Patent # US 9,111,228 B2, Filed Oct 29, 2012, Patented Aug. 18, 2015.

Task 1: Using SAS Studio Programming – Ensemble Segmentation Analytics Example

This next exercise will demonstrate how to use clustering to combine two input segmentations and then use a Bayesian network to adjust any of the input probabilities in the adjustment stage. The data set used is a subset of the Customers data set. The two segmentations are a behavioral cluster segmentation based on similar attributes that we did in Chapter 3 and a survey segmentation that was run on this set of customers asking a number of questions and then assigning a segment based on their survey responses. The goal of the exercise is to combine these two segmentations into a single one with that has good segmentation properties (for example, it won't have a segment with too few or far too many observations). The profile of the combined segmentation depicts features that ensembles the survey and behavioral segments. The brief description of this task is shown in Table 5.1.

Table 5.1: Ensemble Segmentation Steps

Step Number	Brief Process Step Description	Brief Rationale
1	Load the data set ENSEMBLE_DATA into memory and open SAS Studio for coding input.	Ensure that the data set is available for processing.
2	Copy ENSEMBLE_DATA to your casuser library as your local CAS temporary library.	Make a local in-memory copy of the data set for your computations.
3	Perform a k-means clustering using the KCLUS procedure of the two input segmentations. Show a frequency distribution of the newly developed combined cluster segments _cluster_id_.	Use the k-means clustering procedure to perform a combination of the two input segmentations behavior_seg and survey_segments.
4	Run a Bayesian network using the BNET procedure setting the _cluster_id_ as the target and the two input segments as input variables.	Evaluate the combined segmentation using the BNET procedure that can perform Bayesian analytics.
5	Set up a data set for evaluating the probabilities of the cluster segments and compare to the k-means segments. Note differences if any.	Use a fixed probability evaluation to test each of the combined and predicted segmentation probabilities to see whether any adjustments might be needed.
6	Promote the BAYESIAN_EVAL data set for profiling and visualizations.	Copy the data set BAYESIAN_EVAL to a permanent data set and promote it to be in-memory for SAS Visual Analytics exploring and/or reporting.
7	Use SAS Visual Analytics to visualize the output of the ensemble segmentation results.	Build a custom segmentation profile in SAS Visual Analytics and design your own report pages.

Steps 1 and 2

Using the data set ENSEMBLE_DATA provided in the Chapter 5 folder, load this data set into CAS memory so that it can be used for analytic processing.

```
cas mySession sessopts=(caslib=casuser timeout=2400 locale="en_US"
metrics='true');

caslib _all_ assign;
```

```
data casuser.ensemble_data;

  set racoll.ensemble_data(rename=(_segment_=behavior_seg));

drop _segment_label_ distance;

run;
```

Step 3

Now we need to perform a k-means clustering on the two input segmentations: behavior_seg and the survey_segments. The code for this is given below.

```
/* Now run a k-means cluster analysis of the two input segmentations */

proc kclus data=casuser.ensemble_data init=forgy maxiter=50 maxclusters=7
distancenom=globalfreq

      outstat(outiter)=casuser.kclus_stats seed=3712 ;

   input behavior_seg survey_segments /level=nominal;

   output out=casuser.cluster_output copyvars=(_all_);

run;

title 'Distribution of Cluster Frequencies';

proc freq data=casuser.cluster_output;

tables _cluster_id_ ; run; title;
```

Notice that although the two input segmentations are numeric values, we classify them as nominal. This is because k-means treats numeric versus nominal variables very differently. The algorithm used to compute distances are binary, global frequency, and relative frequency. The default, binary, calculates the distance based on simple matching. Global frequency computes the distance based on the frequency count of nominal input variables in the entire data set. Relative frequency computes the distance based only on the nominal inputs for each cluster (SAS Institute, "KCLUS Procedure" 2019). Specifying a seed value ensures that when you re-run the cluster analysis, you will get the same basic results each time. To review the frequency counts of each cluster, the FREQ procedure is used to display that tabular result. The partial outputs of the KCLUS procedure and FREQ procedure are shown below in Figures 5.3a and 5.3b respectively.

Step 4

Now we come to the Bayesian network analysis where we will predict the probability of the newly formed ensemble cluster from k-means using a Bayesian network algorithm. The STRUCTURE statement guides the Bayesian analysis to various network arrangements. There are five different structure argument levels.

Figure 5.3a: KCLUS Procedure Partial Output

The KCLUS Procedure

Number of Observations Read	39109
Number of Observations Used	39109

Model Information	
Clustering Algorithm	K-modes
Maximum Iterations	50
Stop Criterion	Cluster Change
Stop Criterion Value	0
Clusters	7
Initialization	Forgy
Seed	3712
Distance for Nominal Variables	GlobalFreq
Nominal Imputation	None

Frequencies for Nominal Variables

Variable	Level	FrequencyRead	1	2	3	4	5	6	7
behavior_seg	1	2997	0	0	0	0	0	0	2997
	2	2684	0	2679	2	0	3	0	0
	3	9685	0	0	8187	1496	2	0	0
	4	8720	0	0	0	1327	658	6735	0
	5	6686	0	5337	0	961	388	0	0
	6	8337	5950	0	0	1355	1032	0	0
survey_segments	1	2083	0	0	0	0	2083	0	0
	2	7620	616	1317	1279	0	0	1411	2997
	3	5139	0	0	0	5139	0	0	0
	4	13609	4514	504	5890	0	0	2701	0
	5	10658	820	6195	1020	0	0	2623	0

Cluster Summary for Nominal Variables

Cluster	Frequency	Distance from Cluster Centroid to Observation			Within Cluster Distance	Nearest Cluster	Distance to Nearest Cluster Centroid
		Minimum	Maximum	Average			
1	15118	0	2.0000	1.1086	16760.0	3	2.0000
2	7733	0	1.0000	0.6608	5110.0	5	1.0000
3	7171	0	1.0000	0.1789	1283.0	7	1.0000
4	5221	0	1.0000	0.3266	1705.0	3	1.0000
5	2701	0	0	0	0	3	1.0000
6	504	0	0	0	0	4	1.0000
7	661	0	1.0000	0.00454	3.0000	3	1.0000

(Continued)

Figure 5.3b: FREQ Procedure Partial Output

Distribution of Cluster Frequencies

The FREQ Procedure

_CLUSTER_ID_	Frequency	Percent	Cumulative Frequency	Cumulative Percent
1	5950	15.21	5950	15.21
2	8016	20.50	13966	35.71
3	8189	20.94	22155	56.65
4	5139	13.14	27294	69.79
5	2083	5.33	29377	75.12
6	6735	17.22	36112	92.34
7	2997	7.66	39109	100.00

```
title 'General Bayesian Network Assessment of the k-Means Clustering
Combination';

proc bnet data=casuser.cluster_output bestmodel missingnom=level
structure=GN;

   target _cluster_id_ ;

input behavior_seg survey_segments /level=nominal;

output out=casuser.bayesian_results copyvars=(_all_);

run; title;
```

In the SAS VDMML documentation (SAS Institute, "BNET Procedure" 2019), these levels are given, and you select multiple values in the STRUCTURE= statement (separated by commas), and then use the BESTMODEL option. The procedure will select the best model fit with the best structure. In our case, we're using the general Bayesian network structure. The partial output of the Bayesian network is shown in Figure 5.4 below.

Step 5

```
/* Evaluation of Bayesian probability levels and k-Means Cluster Segments
*/

data casuser.bayesian_eval;

   set casuser.bayesian_results;

   if p__cluster_id_1 >= 0.75 then bcluster_id = 1;

      else bcluster_id = i__cluster_id_ ;

   if p__cluster_id_2 >=0.75  then bcluster_id = 2;

         bcluster_id = i__cluster_id_ ;

    if p__cluster_id_3 >= 0.75 then bcluster_id = 3;

         else bcluster_id = i__cluster_id_ ;
```

```
        if p__cluster_id_4 >= 0.75 then bcluster_id = 4;
            else bcluster_id = i__cluster_id_ ;
        if p__cluster_id_5 >= 0.75 then bcluster_id = 5;
            else bcluster_id = i__cluster_id_ ;
        if p__cluster_id_6 >=0.75  then bcluster_id = 6;
            else bcluster_id = i__cluster_id_ ;
        if p__cluster_id_7 >= 0.75 then bcluster_id = 7;
            else bcluster_id = i__cluster_id_ ;
run;
title 'Eval of Bayesian vs. k-Means Cluster IDs';
proc freq data=casuser.bayesian_eval;
    tables bcluster_id * _cluster_id_ ;
run; title;
```

To evaluate how the clustering procedure and the Bayesian network procedure compare, the crosstabulation chart in Figure 5.5 below shows the differences in the Bayesian adjustments from the k-means procedure. The code below gives the Bayesian evaluation and the crosstab from that analysis. At each k-means cluster level, we evaluate the Bayesian probability of each cluster at 0.75 or greater, and if less, then the Bayesian cluster will be the level of the predicted value of the I__CLUSTER_ID_ variable (that is, into the _CLUSTER_ID_ value or the estimated cluster ID).

Step 6

Now we can promote the final resulting data set so that we can explore the results using SAS Visual Analytics. Instead of the OUTCASLIB='racoll' option, you will need to provide your own output library.

```
/* Promote Bayesian Eval data set and save to permanent CAS library RACOLL */.
proc casutil sessref=mySession;
    promote casdata='bayesian_eval' incaslib='casuser'
casout='bayesian_eval'
        outcaslib='racoll';
run;
```

Figure 5.4: Partial Output of the Bayesian Network for Ensemble Predictions

General Bayesian Network Assessment of the k-Means Clustering Combination

The BNET Procedure

Model Information

Significance Threshold	0.05
Prescreening	1
Variable Selection	0
Structure	General
Parenting Method	BestSet
Maximum Number of Parents	2
Missing Interval Variable Handling	Ignore
Missing Nominal Variable Handling	Level
Number of Bins	5
Independence Test	ChiGSquare

Number of Observations

Number of Observations Read	39109
Number of Observations Used	39109
Number of Observations Used for Training	39109
Number of Observations Used for Assessment	39109

Fit Statistics

Number of Nodes	3
Number of Links	3
Average Degree	2
Maximum Number of Parents in Network	2
Number of Parameters	239
Score	-110769.37

Variable Selection

Variable	Selected	Chi-Square	Pr > ChiSq	G-Square	Pr > GSq	Mutual Information	DF	Conditional Variables
behavior_seg	Yes	163587	<.0001	108791	<.0001	0.96854	35	
survey_segments	Yes	68464	<.0001	60547	<.0001	0.88733	28	

Output CAS Tables

Step 7

The CASUTIL procedure allows us to promote the data set from the CAS library and save it in a permanent library (in my case, the library is called RACOLL). This automatically places the data into memory so that we can immediately explore and visualize it with SAS Visual Analytics. In this step, we will take several sub-steps to start our visualization and profiling process.

Step 7a

After the CASUTIL procedure executes the data set, Bayesian_Eval is now saved in a permanent library and also loaded into active memory. Open the upper left menu called **Analytics Life Cycle** and select the option **Explore and Visualize** as shown in Figure 5.6. Select **Start with Data** and

Figure 5.5: Crosstab of the k-Means and Bayesian Network Segmentations

Eval of Bayesian vs. k-Means Cluster IDs

The FREQ Procedure

Table of bcluster_id by _CLUSTER_ID_								
				_CLUSTER_ID_				
bcluster_id	1	2	3	4	5	6	7	Total
1	5950	0	0	0	0	0	0	5950
	15.21	0.00	0.00	0.00	0.00	0.00	0.00	15.21
	100.00	0.00	0.00	0.00	0.00	0.00	0.00	
	100.00	0.00	0.00	0.00	0.00	0.00	0.00	
2	0	8016	0	0	0	0	0	8016
	0.00	20.50	0.00	0.00	0.00	0.00	0.00	20.50
	0.00	100.00	0.00	0.00	0.00	0.00	0.00	
	0.00	100.00	0.00	0.00	0.00	0.00	0.00	
3	0	0	8189	0	0	0	0	8189
	0.00	0.00	20.94	0.00	0.00	0.00	0.00	20.94
	0.00	0.00	100.00	0.00	0.00	0.00	0.00	
	0.00	0.00	100.00	0.00	0.00	0.00	0.00	
4	0	0	0	5139	0	0	0	5139
	0.00	0.00	0.00	13.14	0.00	0.00	0.00	13.14
	0.00	0.00	0.00	100.00	0.00	0.00	0.00	
	0.00	0.00	0.00	100.00	0.00	0.00	0.00	
5	0	0	0	0	2083	0	0	2083
	0.00	0.00	0.00	0.00	5.33	0.00	0.00	5.33
	0.00	0.00	0.00	0.00	100.00	0.00	0.00	
	0.00	0.00	0.00	0.00	100.00	0.00	0.00	
6	0	0	0	0	0	6735	0	6735
	0.00	0.00	0.00	0.00	0.00	17.22	0.00	17.22
	0.00	0.00	0.00	0.00	0.00	100.00	0.00	
	0.00	0.00	0.00	0.00	0.00	100.00	0.00	
7	0	0	0	0	0	0	2997	2997
	0.00	0.00	0.00	0.00	0.00	0.00	7.66	7.66
	0.00	0.00	0.00	0.00	0.00	0.00	100.00	
	0.00	0.00	0.00	0.00	0.00	0.00	100.00	
Total	5950	8016	8189	5139	2083	6735	2997	39109
	15.21	20.50	20.94	13.14	5.33	17.22	7.66	100.00

then select the data set **Bayesian_Eval.** On Page1 of our new profiling report, select the object called **Heat Map** as in Figure 5.7. On the right-hand side in the **Data Roles** icon, select the following categories to the Axis Items below along with the _DISTANCE_ variable for color.

Step 7b

Now create a Page2 of your report and select a **Box Plot** object and drag it to the page. Select the bottom of the box plot object and move the plot object half-way to the middle of the

Figure 5.6: Selecting the Bayesian_Eval Data Set for a Visualization Report

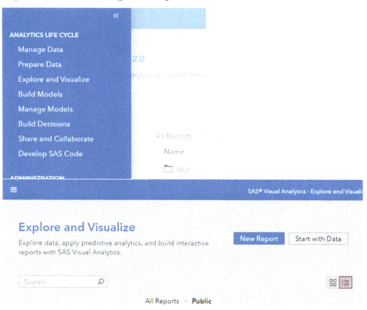

Figure 5.7: Ensemble Segments Profile Visual Report

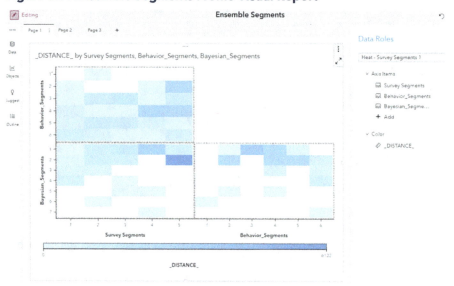

screen. Drag a second Box Plot object on the bottom. On the top box plot, you can select **Survey Segments** in the Category in **Data Roles**, **Log(Employee @ Site)**, and **Number of Yrs. Purchased as Measures**. In the bottom Box Plot object, use the same measures at the top object except select **EM_Eval_Segment** as the Category. This is shown in Figure 5.8 below. You can add a Page3

Figure 5.8: Ensemble Segments Profile Visual Report (Page2)

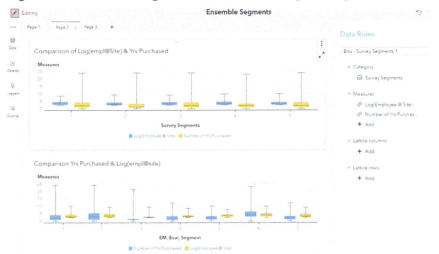

with the same categories and other measures in your Box Plot objects as well as many other visual objects in which to customize your profile report. In this fashion, you can build a detailed and custom cluster segment profile report for the consumers of your segmentation analytics.

Task 2: Build Your Own Ensemble Segmentation

In this next task, try taking the same steps that you did in Task 1 and see how it comes out to have three input segmentations; use **RFM_cell** (recency, frequency, and monetary value segments A-K), **Survey Segments (1-5)**, and **channel_purchase**. One note here: when you perform your evaluation data set, you might need to actually adjust the k-means segments to the inputting Bayesian segments. (Hint: use the I__CLUSTER_ID_ variable as the revised segment when the evaluation of predicted probability isn't greater or equal to 0.75 value.) The code I used is given in the Chapter 5 folder as "My Bayesian Solution.sas", and the partial output showing the difference between the k-means solution and the Bayesian network values is given in Figure 5.9 below.

Let's see what we accomplished in these tasks. In Task 1, the k-means clustering fit the exact same clusters as the predicted Bayesian network did with two observations that didn't match. In Task 2, I suggested that you use the I__CLUSTER_ID_ variable because it is the recorded predicted value of the Bayesian analysis whereas the _CLUSTER_ID_ is the k-means solution result. This overall analysis uses both unsupervised and supervised analytical techniques to accomplish the final resulting ensemble segmentation.

Figure 5.9: Author's Version of "My Bayesian Ensemble" Segmentation

My Eval of Bayesian Ensemble - 3 Input Segmentations

The FREQ Procedure

bcluster_id	Frequency	Percent	Cumulative Frequency	Cumulative Percent
1	4886	12.49	4886	12.49
2	8350	21.35	13236	33.84
3	3664	9.37	16900	43.21
4	4530	11.58	21430	54.80
5	2241	5.73	23671	60.53
6	555	1.42	24226	61.94
7	5732	14.66	29958	76.60
8	1732	4.43	31690	81.03
9	1462	3.74	33152	84.77
10	924	2.36	34076	87.13
11	2058	5.26	36134	92.39
12	2975	7.61	39109	100.00

_CLUSTER_ID_	Frequency	Percent	Cumulative Frequency	Cumulative Percent
1	4950	12.66	4950	12.66
2	8306	21.24	13256	33.90
3	3597	9.20	16853	43.09
4	4530	11.58	21383	54.68
5	2278	5.82	23661	60.50
6	550	1.41	24211	61.91
7	5732	14.66	29943	76.56
8	1732	4.43	31675	80.99
9	1462	3.74	33137	84.73
10	923	2.36	34060	87.09
11	2065	5.28	36125	92.37
12	2984	7.63	39109	100.00

References

Berk, Richardi. A. 2005. "An Introduction to Ensemble Methods for Data Analysis," Department of Statistics, UCLA.

Breiman, Leo. 1996. "Bagging Predictors," Machine Learning, vol. 24 (2), pp. 123-140.

Chen, Tianqi, and Carlos Guestrin. 2016. XGBoost: A Scalable Tree Boosting System. In KDD '16: Proceedings of the 22nd ACM SIGKDD International Conference on Knowledge Discovery and Data Mining (pp. 785-794). New York, NY, USA: ACM. https://doi.org/10.1145/2939672.2939785

Domingos, P. and D. Lowd. 2005. "Naïve Bayes Models for Probability Estimation," Proceedings of the 22nd International Conf. on Machine Learning, Bonn, Germany.

Ghaemi, Reza., Md. Nasir Sulaiman, Hamidah Ibrahim, and Norwati Mustapha. 2009. "A Survey: Clustering Ensemble Techniques," World Academy of Science, Engineering & Technology., vol. 30 (2) , pages 365-374.

Mitchell, Tom M. 1997. "Machine Learning", International Edition, New York: McGraw-Hill, Chapter 6.

SAS Institute Inc. 2019. "The KCLUS Procedure." SAS® Visual Statistics 8.5: Procedures. Cary, NC: SAS Institute Inc. https://go.documentation.sas.com/?cdcId=pgmsascdc&cdcVersion=9.4_3.5&docsetId=casstat&docsetTarget=casstat_kclus_syntax01.htm&locale=en

SAS Institute Inc. 2019. "The BNET Procedure." SAS® Visual Data Mining and Machine Learning 8.5: Procedures. Cary, NC: SAS Institute Inc. https://go.documentation.sas.com/?cdcId=pgmsascdc&cdcVersion=9.4_3.5&docsetId=casml&docsetTarget=casml_bnet_syntax01.htm&locale=en#casml.bnet.proc_structure

Strehl, Alexander., and Joydeep. Ghosh. 2002. "Cluster Ensembles – A Knowledge Reuse Framework for Combining Partitionings," American Association for Artificial Intelligence.

Topchy, A. P., M.H.C. Law, A.K. Jain, and A.L. Fred. 2004. "Analysis of Consensus Partition in Cluster Ensemble," Proceedings of the 4th IEEE International Conference on Data Mining (ICDM 2004), pp. 225-232.

Chapter 6: Tying it All Together: Business Applications

Introduction

This chapter brings together the applications discussed in previous chapters and introduces some areas of clustering and segmentation where additional data manipulation is needed prior to any segmentation. These methods are all available in SAS Viya; however, not all of them are available in the GUI interfaces of SAS Model Studio, SAS Visual Analytics, or SAS Visual Statistics. Some are in procedures or action sets that are easily used as you have seen in the examples earlier in this book. The first area is network analytics, which allows measurements of interconnected items. Items could be social contacts, chats or calls, telecommunication network nodes, and many other types of interconnected networks. Once the metrics are computed in these networks, other analytics such as clustering, predictive models, and data discovery are easily implemented. Next is the notion of clustering time series data that could be from raw transactions or other data that is ordered as a time series. Measuring the distance of each time series as a pattern to other time series will allow the set of time series to be clustered, segmented, and used in other modeling methods. Last is a discussion of market research surveys, which often end up as a segmentation profile from the research report. In this case, the ability to extend the research report from a survey to a larger population base (often called *out-of-sample modeling*) can be a method of predicting segments in a database, customer data store, or a group of patients from a survey profile. These methods are all introduced in this chapter.

Tale of Two Customers

As an example, let's say we have two customers, Jane and Michael. They are a married couple. The vending company for Jane and Michael is a telco operator where Jane and Michael have their account. Jane works from home and uses her mobile phone a lot while she works. If Michael has a conference call on his mobile phone, he does this in his car as he is heading into work each morning. Michael does make other calls too, but they are short and somewhat routine. The minutes and data usage are fairly similar for both Jane and Michael; however, their calling patterns are quite different. Recording the minutes and data usage for Jane and Michael and using those metrics in a cluster segmentation model probably won't pick up the call pattern differences between Michael and Jane. If we were to inspect the calling patterns of Jane compared to Michael, we might find something depicted like Figure 6.1 below.

Metrics such as total calls or SMS text messages counts and recency might not capture the pattern differences between Jane and Michael. Another set of metrics that is designed for network relationships is in order in this case. *Network analytics* is a branch of analytics that deals with diverse types of networks such as electrical networks, telecommunications networks, social networks, transportation networks, and many others as well (Backiel et al. 2016; Ahuja et al. 1993; and Pinheiro 2011). By understanding how Michael and Jane make their calls and SMS text message counts and recency, new metrics can be derived that capture the *strength* and *influence* in each of these networks. The analysis of the network metrics reveals the main characteristics of the social network and makes it possible to understand it from a business point of view. The theory of these networks stems from graph theory (Ahuja et al. 1993), a branch of mathematics that makes it possible to determine certain behaviors in networks. For example, in Jane's network, the green nodes are more randomly spread out; whereas in Michael's network, the green nodes are all clustered. The cluster of green nodes in Michael's network suggest that his calls and SMS text messages to the green nodes I and J are perhaps a group of individuals in the same business and that network is more homophilic or clustered due to the nature of connections. Measuring the degree of connections and clustering as direct rather than using

Figure 6.1: Basic Calling Patterns between Jane and Michael

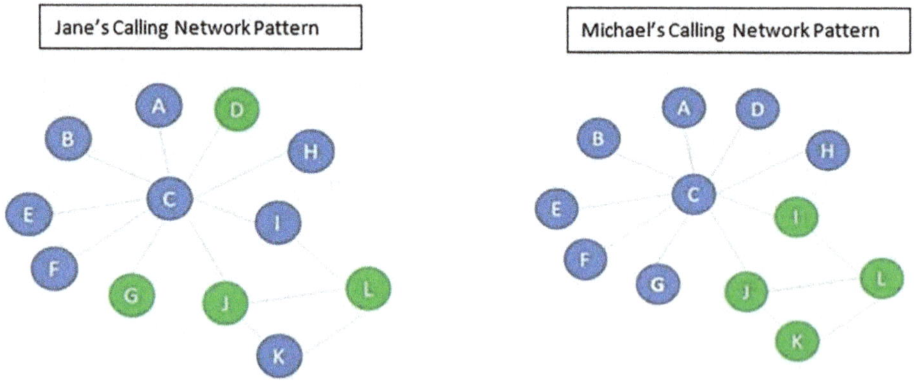

Figure 6.2: Network Versus Customer Attribute Improvements in Churn Model Prediction

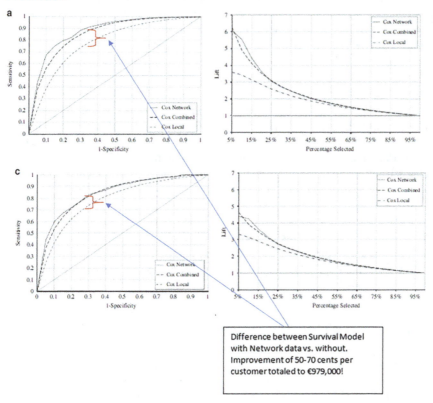

Difference between Survival Model with Network data vs. without. Improvement of 50-70 cents per customer totaled to €979,000!

second- or third-degree connections (indirect connections) allows behavioral characteristics to be captured and used in analytical models.

When these metrics and types of network architectures and structures are captured, they can improve analytical models considerably. A recent study in pre-paid cellular networks shows that the analytics of customer churn was improved greatly using these methods, and the models improved greatly to help save thousands of euros as shown in Figure 6.2 (Backiel et al. 2016).

As you can see, ordinary metrics might not catch certain behaviors. However, once new metrics are developed, then segmentation can take on another dimension.

Applications of Ensemble Segmentation

Another application where segmentation can be assisted is ensemble segmentation as discussed in Chapter 5. One of the more common applications of an ensemble segmentation is when the analyst does not have any of the input attributes that were used to define one or more of the input segmentations. This arises in cases where an organization might send out its customers or prospects to a vendor that would place a segmentation on the sent listing, and then the organization now would have a desired segmentation on its customers or prospects. Since the inputs defining that segmentation are typically not included in such services, the combination of that segmentation would have to be performed by one of the methods described in Chapter 5 (Collica 2015). Combining individual attributes with a segmentation that already has many attributes used to define it from clustering or other means is not likely the best practice. For example, RFM (recency, frequency, and monetary value) is a segmentation of its own; however, it is focused on three attributes of purchase transaction behavior. Contrast RFM with a segmentation that has attributes of both behavior (where RFM might already be included) and demographic. Combining segmentations where similar attributes are defined in both is not a best practice as there is overlap of attributes and the resulting segmentation is difficult to determine whether overlapping attributes existed as there is no correlation history or covariance to account for such inclusion.

Another application might be a micro-segmentation where several input segmentations are combined so that many offer and messaging combinations are devised in marketing communications or in investigative profiling. These ensemble methods allow a data driven approach; however, one could always include an input segmentation purely based on business rules if that is needed for the business application.

Applications of Time Series Segmentations

This section discusses how to capture pattern differences in times series such as transactional data and measure the difference between one time series and other times series thus giving a single metric of those differences. With a new metric of distance between various time series, clustering and segmentation of time series and transactional data captured over time can be accomplished.

There are many applications for time series clustering and other analyses in business and science. Some examples are listed here, but the following is certainly not an exhaustive list.

- Using known fraudulent transactions as the target and finding similar transactions that have a high potential for being fraudulent.
- Finding store revenue transactions to find similarities and differences in a chain of stores.
- Customer purchase transactions of quantities or revenues grouped in discrete time intervals such as weekly, monthly, quarterly, and so on.
- Event history data aggregated in discrete units of time. Any type of customer or prospect event taken over time.
- Online advertising impressions over time aggregated by hour or daypart (Wikipedia) within a week after specific offers, and so on.
- Finding similar patterns in streaming data and clustering to understand key pattern groups.
- Grouping stock exchange transactions and returns over time for pattern similarities and differences.

The ability to process transactions into a times series and measure how similar they are in both magnitude and in time sequence simultaneously is a key capability that should not be overlooked and should prove very valuable in an arsenal of analytics for every organization.

Measuring Transactions as a Time Series

"If time be of all things the most precious, wasting time must be the greatest prodigality." Benjamin Franklin coined that phrase. So how can we best define a time series? Essentially, a *time series* is a collection of observations ordered sequentially in time. Figure 6.3 shows a representation of a typical time series.

The human eye and brain can detect very subtle changes in images. Take, for example, the time series charts in Figures 6.4a and 6.4b. Upon inspection, it is fairly easy to see that these two charts, although similar, are quite different from each other.

In order to segment the time series in Figures 6.3 and 6.4a-b, a method to measure these patterns is needed to measure the distance between a metric or set of metrics that represents changes in patterns over time. In the late 1700s, a French mathematician and physicist named Jean Baptiste Joseph Fourier (1768–1830) contributed the idea of representing a times series as a linear combination of sines and cosines, but keeping the first $n/2$ coefficients. This is now known as the famous Fourier Series and Fourier Transforms as shown in Figure 6.5 below.

The basic idea that Fourier provides in his transform is to reduce the data in the time series to a number or set of numbers that represents the salient features contained in the time series

Figure 6.3: A Typical Times Series

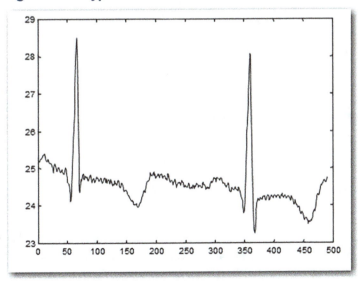

Figure 6.4a: Example 1 Time Series

Response

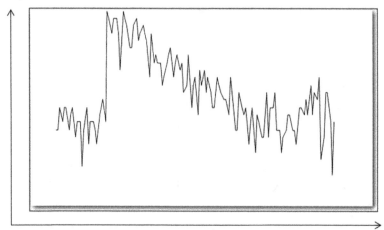

Time Sequence

Figure 6.4b: Example 2 Time Series

Response

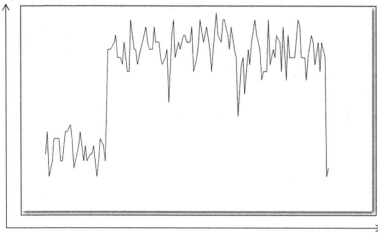

Time Sequence

pattern. By performing a data reduction technique such as this or with another algorithm, you can then perform subsequent analyses with the representative form of the series.

Let's look at another method to represent the time series data. If we would like to use a target series, say the one represented in Figure 6.4a, and compare that with another like in Figure 6.4b and in all other series in our data set, then the metric of similarity could be used in clustering and

Figure 6.5: Fourier Transform for a Time Series

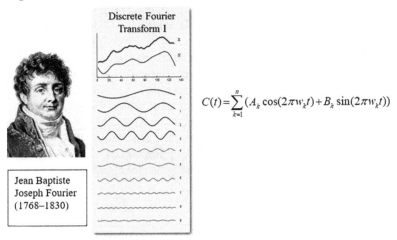

$$C(t) = \sum_{k=1}^{n} (A_k \cos(2\pi w_k t) + B_k \sin(2\pi w_k t))$$

Jean Baptiste
Joseph Fourier
(1768–1830)

segmentation as we have discussed in the previous chapters. One convenient technique is to plot the target sequence with respect to the input sequence. A 45-degree line from the lower left corner to the upper right corner would be an exact match. For each unit of the time sequence, we can measure how different the magnitude and time dimensions are and record each measurement. Such a representation is shown in Figure 6.6.

Figure 6.6: Comparing and Measuring Two Time Sequences

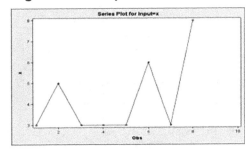

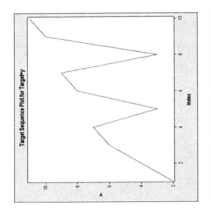

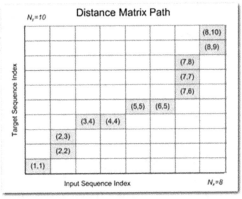

The gray squares represent a coordinate measure for each of the eight-time blocks of the time series sequence. In Figure 6.6 in the bottom left grayed square, the value of (1,1) means that in the first time sequence, the target and the input sequence are exactly the same. In the third time block, the measurement (3,4) depicts that the target has a magnitude of 3, the input has a magnitude of 4, and so on, for each of the eight time periods. Distance metrics can be used in the coordinate pairs in sequence from 1 to 8, representing the magnitude measurements and the time sequence measurements, respectively. SAS Visual Forecasting has provided several object-oriented callable packages for time series analyses (SAS Institute 2019). One of the packages is called the time series distance measure (TSD) package. Each package contains one or more objects that contains methods for analysis that can be called in a SAS program, Jupyter Notebook, or Python or R Studio API call. The following SAS Studio program shown in Program 6.1 takes a data set with columns of counts by sequence and does such measurements.

Program 6.1: Time Series Analysis in SAS Studio

```
/* Start CAS Session on SAS Viya */

cas mySession sessopts=(caslib=casuser timeout=2400 locale="en_US"
metrics='true');

caslib _all_ assign;

data casuser.applianc;

  set sashelp.applianc;

run;

proc sgplot data=casuser.applianc;

title 'Units_5 over Cycle';

 series x=cycle y=units_5;

run;

 /* Run time series similarity on customer account level units purchased
and set up a distance matrix for time series clustering.  */

 proc tsmodel data=casuser.applianc outlog=casuser.outlog

                               outobj=(of= casuser.out_ts_dist(re-
place=yes));

var units_1 units_2 units_3 units_4 units_5 units_6 units_7 units_8
units_9;

id cycle interval=obs;

require tsd;

submit;

    declare object f(DTW);
```

```
    declare object of(OUTTSD);

  rc = f.initialize();

  rc =f.SetTarget(units_1,units_2,units_3,units_4,units_5,units_6,units_7,u
nits_8,units_9);

  rc = f.SetOption("Metric", "Rsqrdev","Normalize","Std","Trim","Both");

  rc = f.Run(); if rc < 0 then stop;

  rc = of.Collect(f); if rc < 0 then stop;

endsubmit;

run;

/* Now cluster the distance metric for the units */

/* convert the distance output data set to a distance matrix for clustering
input */

proc sort data=CASUSER.OUT_TS_DIST out=WORK.sorted;

     by InputSeries;

run;

proc transpose data=work.Sorted out=work.Transpose prefix=Column name=input-
series;

     var Distance;

     by InputSeries;

run;

proc delete data=work.Sorted;

run;

data casuser.transpose; set work.transpose;

run;


/* Cluster Distance Matrix */

ods graphics on;

proc cluster data=casuser.transpose(type=distance) method=ward

          plots=dendrogram(height=rsq);

id inputseries;

run;

ods graphics off;
```

The DATA and SET statements copy the APPLIANC data set from the permanent library called SASHELP into the CASL temporary library called CASUSER. This data set has units 1–9 and will compare all nine with each other. The TSMODEL procedure that can process the time series packages starts with a definition with the variable units 1–9, and the time sequence is the ID that is the variable Cycle, and the time interval is the observation number. The REQUIRE statement lists all the time series packages that will be used in the submit – endsubmit block. In this case, only the TSD package is being required. Since these packages are object-oriented, the declaration statements are for the DTW object and the OUTTSD object. The "f" in front of each method request is just a label that you decide but thereafter use to reference those objects in each of the methods therein. The DTW object can execute many different distance functions to measure a time sequence(s) versus others. In this example, the metrics being used are RSQRDEV for root square deviation. To normalize the sequence, the NORMALIZE argument is using the STD or standard deviation, and the TRIM argument is requesting both the beginning and ending of missing values if any exist. In the clustering, we will need to revise the output data set of measurements for hierarchical clustering being done with a SAS/STAT module procedure called PROC CLUSTER. That procedure can take a lower-triangular matrix data set as an input to the clustering. The dendrogram is then produced where R^2 is the key value plotted for each cluster iteration. Figure 6.7 shows the partial outputs from the code above.

Applying Predictive Analytics to Segments

The main reason and purpose of market research surveys is to obtain specific information from a population being surveyed. Typically, an endpoint of the project is to have a written report or presentation to disseminate the information obtained from the research, survey, and associated analytics of the survey responses. While the report or presentation does convey the objectives

Figure 6.7: Time Series Measurement and Clustering Example Output

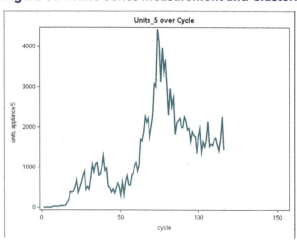

(Continued)

The CLUSTER Procedure
Ward's Minimum Variance Cluster Analysis

Root-Mean-Square Distance Between Observations 4.370909

Cluster History

Number of Clusters	Clusters Joined		Freq	Semipartial R-Square	R-Square	Tie
7	units_4	units_9	2	0.0012	.999	
6	units_3	units_6	2	0.0472	.952	
5	units_2	CL6	3	0.0565	.895	
4	CL7	units_7	3	0.0838	.811	
3	CL5	units_8	4	0.1728	.639	
2	CL3	units_5	5	0.2368	.402	
1	CL2	CL4	8	0.4018	.000	

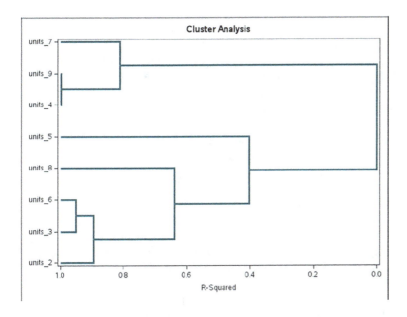

of the study, the study parameters, and the results and conclusions, most market research projects end at this point. This chapter will demonstrate that when designed properly, the market research survey can supply the report or presentation, along with an analytic model that will allow the results of the survey to be extended to the customer or prospect database. The benefit of this approach is that the ROI of the survey is much higher with an analytic model that can extend the results to a customer or prospect database than if only a report is developed and disseminated.

Market research is a broad topic. The area that we are going to focus on is market research a company might use to derive insights on customers or prospects that it cannot obtain in the data it typically has in purchase or sales data, contact information, or sales operational systems. Market and survey researchers gather information about what people think on a topic of interest. Market analysts who analyze surveys help companies understand what types of products people like and want, determine who will buy them, and the price that they are willing to pay for such products or services. Research analysts formulate methods and processes for obtaining the data that they need by designing surveys to assess consumer preferences and choices. Most of these surveys are typically conducted using the Internet or phone. However, sometimes methods such as customer focus group discussions and personal interviews can be used as well. Survey researchers also gather information about people's opinions and attitudes on various issues of interest to organizations, companies, or governments. Survey researchers often focus their design efforts on the set of questions and how to frame these questions to ascertain the desired information about attitudes and opinions of people (Bureau of Labor Statistics 2020). Often, a requirement in their survey design is the target population in which to gather their responses. Responses need to be representative of a larger population if inferences about the population are to be made with the market research analytics.

The majority of market research surveys focuses on the set of questions to ask the survey recipients, the analysis of the survey responses, and the final research report, which summarizes the findings of the research and analytics. However, after the research final report or presentation, most market research ends. This is where I want to describe how to extend the research so that the results of the survey analytics can be applied to a larger set of customers or prospects from which the survey recipients were initially drawn.

Several key items will need to be done during the research design phase of the project in order to ensure that customer or prospect data can be matched back to the database for model development and deployment of scoring results.

Analysis of Survey Responses: An Overview

As said earlier, a typical main goal of market research is to have some insights to help in market planning. However, if all one gets is a report that describes the results, then analytically the report isn't very actionable. If we can extend the research to estimate the results on a larger audience, then that could be much more actionable and useful in addition to the research report! While it is not the intention of this chapter to review the analytic methods of market research surveys, the basics will be discussed. However, the area for specific focus is the resulting segments that are derived from the statistical techniques such as discrete choice modeling, factor or discriminant analysis, maximum difference preference scaling, and covariance analysis. Statistical analysis, with respect to market research methods, can typically be categorized into two basic groups: descriptive statistics and statistical inference. In *descriptive statistics*, basic measures such as mean, variance, frequency counts, and distributions are used to characterize and profile the question responses and perhaps other information gathered from the survey such as company size, or in consumer area items such as household income and other demographic attributes. In *statistical inference*, a key hypothesis or set of hypotheses about the population are tested using sample data. The claims about these hypotheses are what we would like to test; to see whether they are true or false (Bradburn and Sudman 1988).

Predicting Attitudinal Survey Segments

In this example, the survey data was designed to ask enterprises with 1,000 or more employees a series of questions regarding their brand preference, references to purchases, their total company's strategy, and so on. This set of survey responses included a little over 1,000 respondents in the US and Canada. After matching these survey responses to a syndicated B-to-B data set, the data was then expanded from one of the qualifying questions in the survey. One question was asked if their responses applied to the entire company or just the current company site. This was an important feature because approximately 70% of the respondents indicated that their response applied to the entire company! This allowed the response to all the syndicated sites that are linked to the parent company be applied accordingly, while the remaining responses were only at the existing site location. A total of 39,109 customer records had the survey segment applied in total. The task is to see if a predictive model using customer data could predict the five survey segments.

The five survey attitudinal segments were as follows:

1. Trailblazers
2. Adopters
3. Minimalists
4. Self-Starters
5. Conservatives

"Trailblazers" are companies that felt that the use of the latest information technology (IT) products and services greatly enhanced their company's competitiveness and was part of their overall strategy. "Adopters" felt that their company was growing at a rapid rate and therefore could not spend the time to effectively evaluate new IT products and services. They would simply adopt a current technology and use it but not invest in *new* and upcoming technology. "Minimalists" are companies that used IT technology only to keep the lights on and barely made any investments unless something was broken. "Self-Starters" are companies that would not normally purchase IT technology but would generally develop their own in-house technology that they used. "Conservatives" were companies that wanted to invest in new IT technology but needed proof-of-concept in order for them to see and understand the value and incorporate it into their company plans and strategies. The syndicated data purchased contained firmographic information such as industry codes, company size and revenues, geography, addresses, and so on. The column attributes of the data set SURVEY_SEGMENTS are given in Appendix 1. Table 6.1 shows the brief outline of analyzing the survey segments data set.

Task 1: Predicting Survey Segments from a Market Research Survey

Steps 1-3

After loading data into SAS Viya CAS memory for active use, create a VDMML pipeline with the default project settings. Project settings are located in the upper right corner with an icon that looks like a gear ⚙ ▾ . The first tab will always be the Data tab where you set variable attribute settings.

Table 6.1: Steps for Predicting Survey Segments

Step Number	Brief Process Step Description	Brief Rationale
1	Load the data set CUST_SURVEY_SEGMENT into the SAS Viya CAS in-memory.	Ensure that the data set is available for processing.
2	Create a VDMML pipeline in SAS Model Studio.	
3	Set the properties for the data set and reject some attributes.	
4	Create the pipeline with several models.	
5	Add a Forest, Decision Tree, and Bayesian Network node to the pipeline.	
6	Add an Ensemble model and modify the Decision Tree, Forest, and Bayesian Network nodes. Modify the Ensemble node's properties. Re-run the pipeline.	
7	Open the Pipeline Comparison and Insights tabs and review the results.	

Set the following variables to a **Target** and reject some of the variables shown in Figure 6.8. Variables **SIC8**, **SALES_CLASS**, and **STATE** should have a Role of **Rejected**. The variable **SURVEY_SEGMENTS** should have a Role of **Target**.

The SALES_CLASS variable is rejected due to its attributes being already included in another variable called RFM. The SIC8 variable is rejected due to its many categorical levels because these are industry classification codes, and a variable that has aggregated industry verticals is

Figure 6.8: Data Attribute Roles and Settings for Survey Segment Prediction

Predict Survey Segment

Pipelines   Pipeline Comparison   Insights

	Variable Name	Label	Type	Role ↓	Level	!	CUST_SURVEY_SEGMENT
☐	survey_segments	Survey Segment Number Response	Numeric	Target	Nominal		Columns: 48
☐	sales_class	Sales Customer Classification Code	Character	Rejected	Nominal		Rows: 39,109
☐	SIC8	Eight Digit Primary Std. Industry Class Code	Character	Rejected	Nominal		Label: (not available)
☐	STATE	State Customer is Located In	Character	Rejected	Nominal		Location: cas-shared-default/racoll

the INDUSTRY_VERT variable. The STATE variable is also rejected because the REGIONAL_GEO variable has state as its input. The remainder of the variables can remain as Inputs.

Step 4

Figure 6.9: Transformation Node Settings

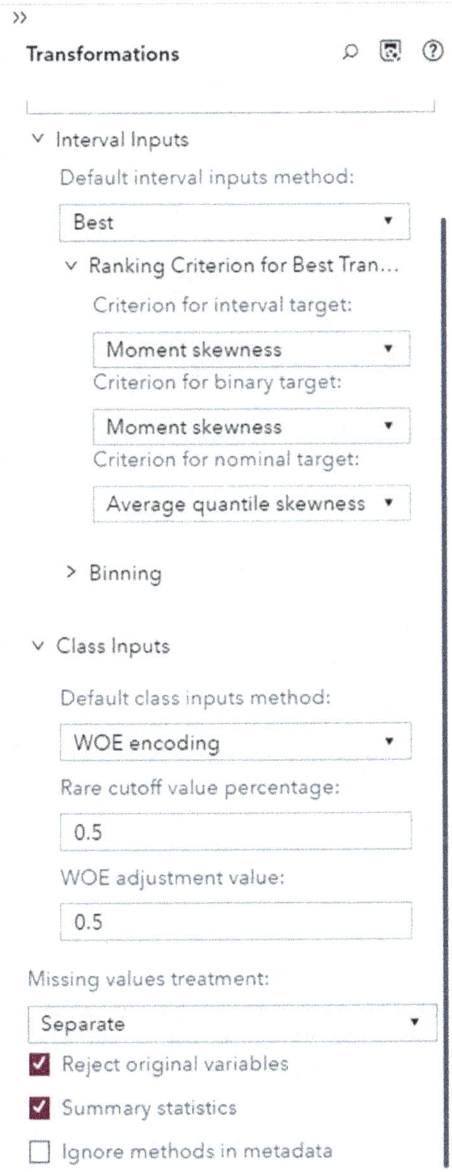

Go to your Pipelines tab. Your newly created pipeline should only have a Data node in it at present. We will need to transform some variables to aid in the modeling process. Place a Transformation node (found in the Data Mining Preprocessing nodes) in the pipeline and set the node settings to the options shown in Figure 6.9.

The BEST setting for interval numeric variables means that the Transformation node will attempt to transform each variable with various transforms and select the best with the strongest association metrics. For classification variables, I selected WOE encoding. WOE is an acronym for weight-of-evidence. WOE is computed by iteratively assigning bins of the classification levels and then testing the WOE adjusted variable for strength of association compared to the original variable. Transform variables, if higher association, will be the variables used in the modeling and subsequent nodes, whereas the original variables will be rejected for use. You can always add a Manage Variables node and change variable roles within the pipeline flow.

Step 5

Now add three modeling nodes to your pipeline after the Transformation node: a **Forest** node (found under the supervised Learning nodes only), **Decision Tree**, and a **Bayesian Network** node. Follow the following settings for each node respectively in Figures 6.10a–d.

In a Random Forest model, I recommend using the Autotuning option to be set because you might or might not know all the best practical options that will provide the optimal result. SAS Model Studio nodes are optimized using the SAS internal optimization engine. Now add a Decision Tree with the settings shown in Figure 6.10b.

Again, I have selected to run Autotuning to optimize the settings. Now add the third modeling node, which is a Bayesian Network model, and use the settings in Figure 6.10c. Note that all settings can be left at their default values.

Now run this pipeline with the three models and then open the Model Comparison node to see which of the three models is the best.

Step 6

Now that we have a winner or champion model, let's try an additional modeling type. It is often true that combination models win out against single models due to behavioral characteristics that one model might detect whereas other might not. Combining the models can at times produce a lower misclassification rate or higher KS or an increase in the ROC (area under the curve) statistic. So add an Ensemble node to the Forest node. Then in the Ensemble node, add other modeling nodes – the Decision Tree and the Bayesian Network. Then, once added we will need to set up how we want the Ensemble node to combine the models. Figure 6.10d shows the Ensemble Node settings that I had selected. In adding an Ensemble node to combine models, you need to right mouse click on one of the modeling nodes, then select **Add Child node**, and then select **Postprocessing**, and then **Ensemble**. Once your Ensemble node has been added, if

Figure 6.10a: Forest Node Settings (Only Change Is to Turn on Autotuning)

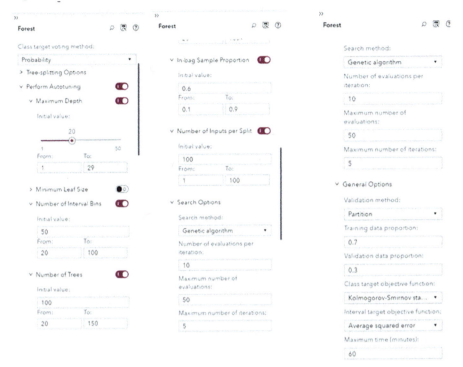

Figure 6.10b: Decision Tree Node Settings (Only Change Is to Turn on Autotuning)

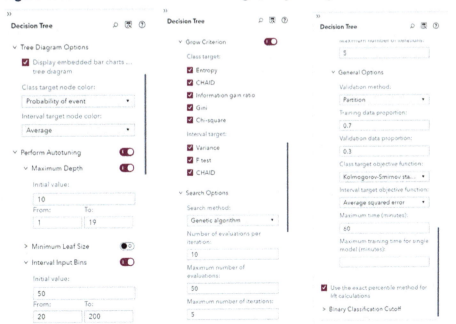

(Continued)

Figure 6.10c: Bayesian Network Node Settings (All the Settings in This Node Can Be the Default Settings)

Figure 6.10d: Ensemble Node Settings

Ensemble Model Settings:

(Continued)

Figure 6.10e: Predicting Survey Segments Pipeline

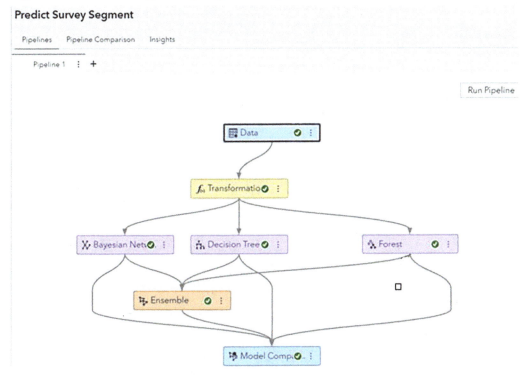

you right-click on the Ensemble node, you can select **Add Models**. Other models already in your pipeline can then be added. Once you have edited the Ensemble node settings, your pipeline should look like that in Figure 6.10e. Figure 6.11 shows the Model Comparison node's results indicating that the Forest node is the champion.

Now we see that the best model is the Random Forest. Add a score node after the Forest node and score all three training, validation, and test data sets and save the output to your permanent CAS library. Figure 6.12 shows the completed pipeline flow.

Step 7

Go to the Pipeline Comparison tab and review the outputs therein. Figures 6.13a and 6.13b show a partial set of graphics in the Pipeline Comparison.

Figure 6.11: Model Comparison Node – Node and Assessment

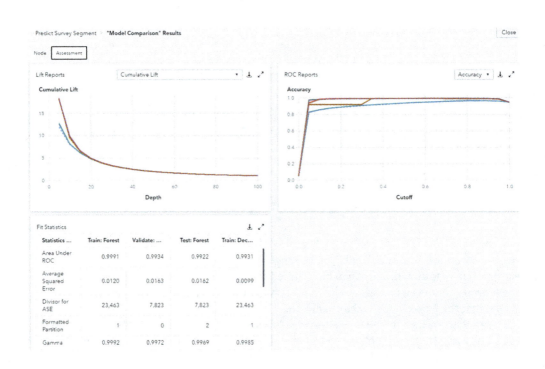

Figure 6.12: Completed Pipeline Flow for Predict Survey Segment Project

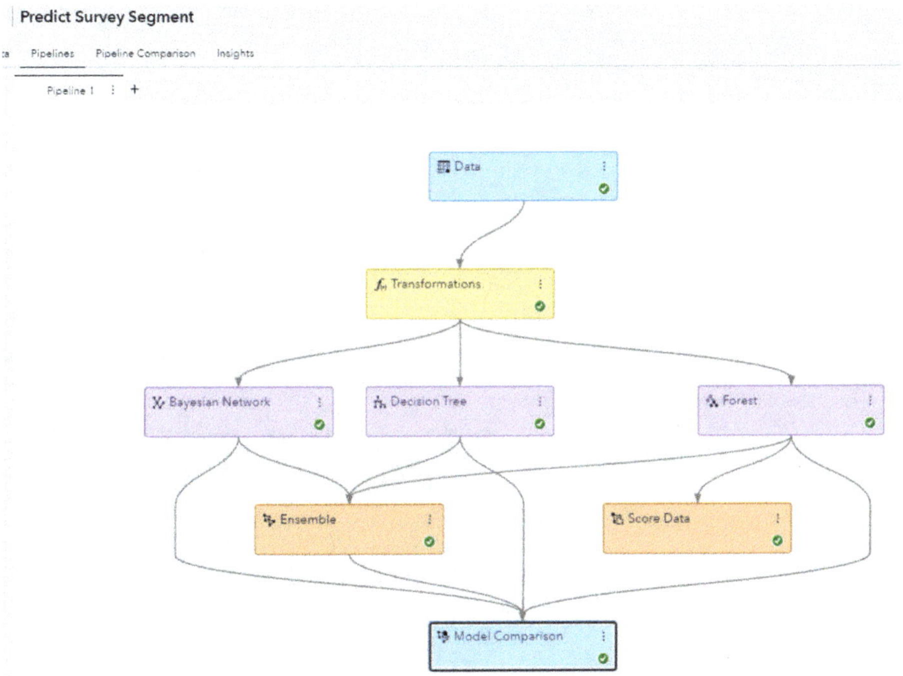

Figure 6.13a: Pipeline Comparison Tab (Partial Output Shown)

(Continued)

Figure 6.13b: Pipeline Comparison Tab

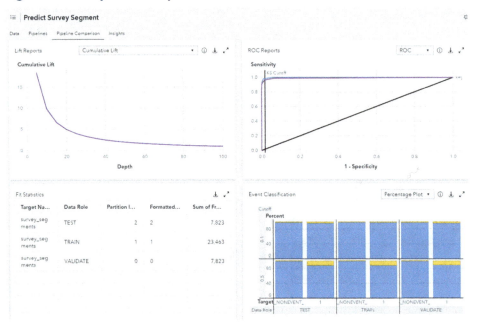

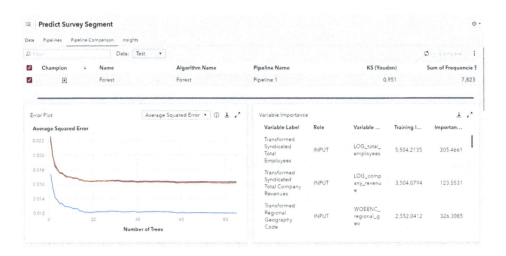

In Figure 6.14, the Insights tab of your pipeline contains the results of your entire process flow explained as a project summary.

Figure 6.14: Insights Tab Results (Partial Output Shown; This Is the Rightmost Tab in the Pipeline)

If you were to review the Insights tab, the top five attributes are Transformed Total Employees, Transformed Total Company Revenues, Transformed Regional Geo Code, Transformed Phone Contact restriction (Y/N), and Transformed Aggregated Industry Vertical Code (Figure 6.15). While other attributes certainly contributed, the top five are obviously the strongest predictors. Since the untransformed variable attributes were rejected, only transformed variables will be used. Sometimes it is good not to reject even a transformed variable as at times some models might prefer the untransformed variables.

This completes the analysis of predicting a survey segment use case. It is my hope that these examples in this and previous chapters have given you good ideas on how to accomplish

Figure 6.15: Top Five Variable Importance from Insights Tab

customer segmentation with a variety of approaches, visualizations, and explorations in SAS Viya using SAS Model Studio, SAS Visual Data Mining and Machine Learning (VDMML), SAS Studio code, SAS Visual Statistics, and SAS Visual Analytics.

References

Ahuja, Ravindra K., Thomas L. Magnanti, and James B. Orlin. 1993. Network Flows: Theory, Algorithms, and Applications. Englewood Cliffs, New Jersey: Prentice Hall.

Backiel, Aimee, Bart Baesens, and Gerda Claeskens. 2016. "Predicting Time-to-Churn of Prepaid Mobile Telephone Customers using Social Network Analysis," Journal of the Operational Research Society, 67(9): 1135-1145.

Bradburn, Norman M., and Seymour Sudman. 1988. Polls and Surveys: Understanding What They Tell Us. San Francisco: Jossey-Bass.

U.S. Bureau of Labor Statistics, U.S. Department of Labor. 2020. Occupational Outlook Handbook, Survey Researchers, at https://www.bls.gov/ooh/life-physical-and-social-science/survey-researchers.htm (visited February 10, 2021).

Collica, Randall S. 2015. "System and Method for Combining Segmentation Data," US Patent, Applicant: SAS Institute Inc., NC, Patent # US 9,111,228 B2, Filed Oct 29, 2012, Patented Aug. 18, 2015.

Reis Pinheiro, Carlos Andre. 2011. Social Network Analysis in Telecommunications. Hoboken, New Jersey: Wiley.

SAS Institute Inc. 2019. SAS® Visual Forecasting 8.5: Time Series Packages. Cary, NC: SAS Institute Inc.

Wikipedia, Dayparting. https://en.wikipedia.org/wiki/Dayparting

Appendix 1: Data Sets and Attribute Descriptions

The three data sets below are used in the examples in this manuscript.

CUSTOMERS Data Set

VARIABLE NAME	LENGTH	LABEL
CITY	28	
PURCHFST	8	Year of 1st Purchase
PURCHLST	8	Last Yr of Purchase
Prod_A	8	
Prod_A_Opt	8	
Prod_B	8	
Prod_B_Opt	8	
Prod_C	8	
Prod_D	8	
Prod_E	8	
Prod_E_Opt	8	
Prod_F	8	
Prod_G	8	
Prod_H	8	
Prod_I	8	
Prod_I_Opt	8	
Prod_J	8	
Prod_J_Opt	8	
Prod_K	8	
Prod_L	8	

(*Continued*)

CUSTOMERS Data Set (*Continued*)

VARIABLE NAME	LENGTH	LABEL
Prod_L_Opt	8	
Prod_M	8	
Prod_N	8	
Prod_O	8	
Prod_O_Opt	8	
Prod_P	8	
Prod_Q	8	
RFM	1	Recency, Freq, & Monetary Value Code
SEG	3	Industry Segment Code
STATE	2	
channel	8	Purchase Sales Channel
corp_rev	8	Corporate Revenue last fiscal yr.
cust_flag	3	y/n
cust_id	9	Customer ID No.
customer	1	A=New Acquisition, C=Churn (no purch), R=Cont-Purchase
est_spend	8	Estimated Product-Service Spend
loc_employee	8	No of local employees
public_sector	8	0-No, 1=Yes
rev_class	1	Revenue Class Code
rev_lastyr	8	Last Years Fiscal Revenue
rev_thisyr	8	This Years Fiscal Revenue YTD
tot_revenue	8	Revenue for All Years
us_region	35	US Region Location of Business
yrs_purchase	8	No of Yrs Purchase

In the CUSTOMERS data set, most of the variables have labels; the exception is product columns. Products A–Q are the count of products purchased for each type (letter designation). If no products were purchased, then an empty value "." is indicated. The Cust_Flag is a Y/N designation, and the customer field is unary. The Channel_Purchase indicator has these designations: 0-none, 1-Reseller Only, 2-Direct Only, 3-Both Direct, and Indirect purchase. The file size is 35Mbytes with 44 columns and 105,465 observations.

ENSEMBLE Data Set

VARIABLE NAME	LENGTH	LABEL
Distance	8	Distance
FY1984	8	Fiscal Yr 1984 Revenues
FY1985	8	Fiscal Yr 1985 Revenues
FY1986	8	Fiscal Yr 1986 Revenues
FY1987	8	Fiscal Yr 1987 Revenues
FY1988	8	Fiscal Yr 1988 Revenues
FY1989	8	Fiscal Yr 1989 Revenues
FY1990	8	Fiscal Yr 1990 Revenues
FY1991	8	Fiscal Yr 1991 Revenues
FY1992	8	Fiscal Yr 1992 Revenues
FY1993	8	Fiscal Yr 1993 Revenues
FY1994	8	Fiscal Yr 1994 Revenues
FY1995	8	Fiscal Yr 1995 Revenues
FY1996	8	Fiscal Yr 1996 Revenues
FY1997	8	Fiscal Yr 1997 Revenues
FY1998	8	Fiscal Yr 1998 Revenues
FY1999	8	Fiscal Yr 1999 Revenues
FY2000	8	Fiscal Yr 2000 Revenues
FY2001	8	Fiscal Yr 2001 Revenues
FY2002	8	Fiscal Yr 2002 Revenues
FY2003	8	Fiscal Yr 2003 Revenues
FY2004	8	Fiscal Yr 2004 Revenues
FY2005	8	Fiscal Yr 2005 Revenues
FY2006	8	Fiscal Yr 2006 Revenues
FY2007	8	Fiscal Yr 2007 Revenues
PWR_company_revenue	8	Transformed: Syndicated Total Company Revenues
RESTRICT_EMAIL	1	If Email Contact Restricted (Y/N)
RESTRICT_MAIL	1	If Direct Mail Restricted (Y/N)
RESTRICT_PHONE	1	If Phone Contact Restricted (Y/N)
SIC8	8	Eight Digit Primary Std. Industry Class Code
STATE	2	State Customer is Located In
TRANS_0	8	Log(IT Spend)
TRANS_1	8	Log(Employee @ Site)

(Continued)

ENSEMBLE Data Set (*Continued*)

VARIABLE NAME	LENGTH	LABEL
TRANS_2	8	Log(Total Employees)
SEGMENT1	8	Segment Id
SEGMENT	8	Segment Id
_SEGMENT_LABEL_	80	Segment Description
channel_purchase	8	Channel Customer Purchased
company_revenue	8	Syndicated Total Company Revenues
cust_site_id	46	Customer Identifier
employee_atsite	8	Syndicated Site No of Employees
first_purch_yr	8	First Yr Customer Purchased
industry_vert	3	Aggregated Industry Vertical Code
it_budget	3	IT Budget Range A-E
it_spending	8	Estimated IT Spending in $
last_purch_yr	8	Last Yr Customer Purchased
regional_geo	3	Regional Geography Code
rfm_cell	1	RFM Cell Code A-K
sales_class	4	Sales Customer Classification Code
survey_segments	8	Survey Segment Number Response
synd_id2	9	Syndicated 2nd Level ID
synd_id3	9	Syndicated 3rd Level ID
synd_id4	9	Syndicated 4th Level ID
tot_rev_allyrs	8	Total Revenue All Years
total_employees	8	Syndicated Total Employees
years_purchased	8	Number of Yrs Purchased

Variables FY1984–FY2007 are fiscal year revenues from customer purchases. IT Budget is the estimated IT spend in categories A–E (in units of thousands; A <= $500, B: $500-$999, C: $1,000-$9,999, D: $10,000-$49,999, E: >= $50,000). RFM is a recency, frequency, monetary value cell that has levels of A–K and the recency is the two most recent fiscal years of purchases. IT Spending is the estimated dollar amount from the IT budget category levels A–E. _Segment_ and _Segment_Label_ are cluster segments based on behavioral data. Survey_Segments are five levels 1–5 of attitudinal segments from a survey, 1 being minimal attitude toward IT products/ services, and 5 being tech-savvy of IT products/services.

CUST SURVEY SEGMENT Data Set

VARIABLE NAME	LENGTH	LABEL
FY1984	8	Fiscal Yr 1984 Revenues
FY1985	8	Fiscal Yr 1985 Revenues
FY1986	8	Fiscal Yr 1986 Revenues
FY1987	8	Fiscal Yr 1987 Revenues
FY1988	8	Fiscal Yr 1988 Revenues
FY1989	8	Fiscal Yr 1989 Revenues
FY1990	8	Fiscal Yr 1990 Revenues
FY1991	8	Fiscal Yr 1991 Revenues
FY1992	8	Fiscal Yr 1992 Revenues
FY1993	8	Fiscal Yr 1993 Revenues
FY1994	8	Fiscal Yr 1994 Revenues
FY1995	8	Fiscal Yr 1995 Revenues
FY1996	8	Fiscal Yr 1996 Revenues
FY1997	8	Fiscal Yr 1997 Revenues
FY1998	8	Fiscal Yr 1998 Revenues
FY1999	8	Fiscal Yr 1999 Revenues
FY2000	8	Fiscal Yr 2000 Revenues
FY2001	8	Fiscal Yr 2001 Revenues
FY2002	8	Fiscal Yr 2002 Revenues
FY2003	8	Fiscal Yr 2003 Revenues
FY2004	8	Fiscal Yr 2004 Revenues
FY2005	8	Fiscal Yr 2005 Revenues
FY2006	8	Fiscal Yr 2006 Revenues
FY2007	8	Fiscal Yr 2007 Revenues
RESTRICT_EMAIL	1	If Email Contact Restricted (Y/N)
RESTRICT_MAIL	1	If Direct Mail Restricted (Y/N)
RESTRICT_PHONE	1	If Phone Contact Restricted (Y/N)
SIC8	8	Eight Digit Primary Std. Industry Class Code
STATE	2	State Customer is Located In
channel_purchase	3	Channel Customer Purchased
company_revenue	8	Syndicated Total Company Revenues
cust_site_id	46	Customer Identifier

(Continued)

CUST SURVEY SEGMENT Data Set (*Continued*)

VARIABLE NAME	LENGTH	LABEL
employee_atsite	8	Syndicated Site No of Employees
first_purch_yr	3	First Yr Customer Purchased
industry_vert	3	Aggregated Industry Vertical Code
it_budget	3	IT Budget Range A-E
it_spending	8	Estimated IT Spending in $
last_purch_yr	3	Last Yr Customer Purchased
regional_geo	3	Regional Geography Code
rfm_cell	1	RFM Cell Code A-K
sales_class	4	Sales Customer Classification Code
survey_segments	8	Survey Segment Number Response
synd_id2	9	Syndicated 2nd Level ID
synd_id3	9	Syndicated 3rd Level ID
synd_id4	9	Syndicated 4th Level ID
tot_rev_allyrs	8	Total Revenue All Years
total_employees	8	Syndicated Total Employees
years_purchased	3	Number of Yrs Purchased

The CUST SURVEY SEGMENT data set is nearly identical to the ENSEMBLE data set. The survey_ segments fields contain the cluster segment from a marketing research survey.

Appendix 2: Review of Clustering

What Is Similar and What Is Not

Sometimes the phrase "look at the big picture" is used to back away from the details and look at what the *main* patterns or effects are telling you about the set of customers, patients, prospects, cancer treatment records, or measures of star temperature and luminosity. When a database contains so many variables and rows or records and so many possibilities of dimensions, the structure could be so complex that even the best set of *directed* data mining techniques are unable to ascertain meaningful patterns from it. I use the term *directed* as most data mining techniques are classified into either *directed* or *undirected.* In directed data mining, the goal is to explain the value of some particular field or variable (income, response, age, and so on) in terms of the other remaining variables available. A target variable is chosen to tell the computer algorithm how to estimate, classify, or predict its value. Undirected data mining does not have a predicted variable; instead, we are asking the computer algorithm to find a set of patterns in the data that might be significant and perhaps useful for the purpose at hand (Berry and Linoff 1997). Another way of thinking about undirected data mining or knowledge discovery is to *recognize* relationships in the data and directed knowledge discovery to *explain* those relationships once they have been detected. In order for an algorithm to find relationships, a set of rules or criteria must be made to measure the associations among the individuals so that patterns can be detected.

One way to think of similarity (or the converse, dissimilarity) is to devise a specific metric, measure the items of interest, and then classify the items according to their measures. For example, a set of students in a class could be measured for their height. Once each student is measured, one could classify the students as tall, medium, or short. A secondary characteristic, sex, could then be added to the height measurement, and now the measure of similarity is both height and sex in combination. This concept is shown in Figure A2.1.

Figure A2.1: Two Measures of Similarity (Height and Gender)

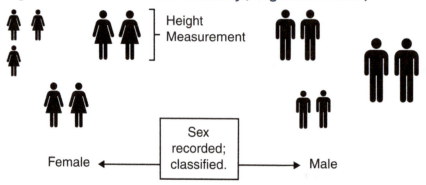

Now that both characteristics are measured or recorded, they can be classified according to a set of criteria. Here, the measure of similarity is the classified sex of an individual and the physical height of the individual. The concept that is taking place in this example is really a measure of *association* between the individual's using two characteristics, sex and height. Notice in Figure A2.1 that the height of the individuals forms three distinct groups: short, medium, and tall heights. It is somewhat intuitive that the three groups all share something in common within each group; the members have similar heights and they are of the same sex. For practical purposes, the definitions of similarity, association, and distance are all considered synonymous. These techniques form the basis of how to measure the distance or association between records of customers or prospects in a database. There are a few caveats, however, that we will need to consider.

Distance Metrics as a Measure of Similarity and Association

How can we measure distance between records in a database on several variables that have different scales and different meanings? To demonstrate this concept a little further, consider a database of attributes as shown in Table A2.1. The fields (columns) in the database have

Table A2.1: Database Field Descriptions with Different Attribute Types

Field Name/Description	Measurement Type	Scale	Units
Last fiscal year revenue	Numeric	Interval	$
Filed a tax return on time (Y/N)	Character	Binary	None
Responded to direct mail (1/0)	Numeric	Binary	None
Year company was founded	Numeric	Ordinal	Years
Credit rating score (1–6)	Numeric	Ordinal	None
Industry group code	Character	Nominal	None
Distance to nearest major metro area	Numeric	Interval	Miles

diverse types (numeric, character) and scales (for example, binary, ordinal, nominal, and interval), and they have different units.

If we were to measure the distance between customer records based on the variables in Table A2.1, what would be the unit of measurement when combining dollars, years, miles, industry code, and yes or no? In addition, a slight change in the last fiscal year's revenue is not the same as a slight change in miles (distance to the nearest major metro). We must translate the general concept of association into some sort of numeric measure that depicts the degree of similarity as measured by distance. The most common method, but not the only one available, is to translate all the fields under consideration into numeric values so that the records can be treated as points in space using a Cartesian coordinate system. This is desirable because the distance between points in space can be measured from basic Euclidean geometry and a little vector algebra. It is the concept that items that are closer together distance-wise are more similar than items that are farther away from each other. This does depend on the type of distance metric used; however, the basic idea of similarity is strongly associated with numeric distance.

Let's take a simple example to demonstrate how distances can be measured on three rows with two fields in a database, one called AGE and one called VALUE. Table A2.2 shows the three records of AGE and VALUE. So, let's construct some distance measurements from this data set in Table A2.2. To compute distances for the AGE variable, each row must be compared with every other row along with itself. The same will be true for the VALUE variable as well. The distance measurements for AGE starting with row 1 are $8 - 8 = 0$, $8 - 3 = 5$, and $8 - 1 = 7$. Now these are compared with the first row as the reference point. If we increment by one row and do the same, we end up with the following for the AGE variable: Row 2: $3 - 3 = 0$, $3 - 1 = 2$, $3 - 8 = -5$ and for the last row, Row 3: $1 - 1 = 0$, $1 - 3 = -2$, and $1 - 8 = -7$. This completes each of the distance measurements for AGE. We can do similar measurements for the VALUE variable as well, but you probably get the idea.

Table A2.2: Simple Three-row Table of Age and Value

Row #	Customer ID	Age	Value
1	Cust_A	8	$18.50
2	Cust_B	3	$3.30
3	Cust_C	1	$9.75

Table A2.3: Simple Distance Matrix for Age

Row #	Cust_A	Cust_B	Cust_C
1 Cust_A	$8 - 8 = 0$	$3 - 8 = -5$	$1 - 8 = -7$
2 Cust_B	$8 - 3 = 5$	$3 - 3 = 0$	$1 - 3 = -2$
3 Cust_C	$8 - 1 = 7$	$3 - 1 = 2$	$1 - 1 = 0$

In SAS 9.4, the DISTANCE procedure can be used to calculate these distances, which are Euclidean distances; however, it can compute many other distance metrics as well.

To plot these distances on a two-dimensional graph, we will need to calculate distances.

Figure A2.2 shows how distances are measured from points in a simple X-Y plane. Points X and Y are two data points. With a little help from linear algebra and geometry, we will now review some of the formulations to measure distances. The distance from point O (the origin) to point B is $|x|\cos\alpha$ and is well known from elementary geometry. This quantity is also known as the orthogonal projection of X onto Y. The points in this plane can be described as a vector from one point to another. In terms of linear algebra, the two vectors X and Y can be described as the inner product (or scalar product) and is given by Equation A2-1. The inner product of a vector with itself has a special meaning and is denoted as X^T X, and is known as the sum of squares for X. The square root of the sum of squares is the Euclidean norm or length of the vector and is written as $|X|$ or $\|X\|$.

$$\langle X,Y \rangle = X^T Y = \sum_{i=1}^{n} x_i y_i \qquad \text{A2-1}$$

With this type of notation, another way of expressing the inner product between X and Y is given by Equation A2-2.

$$X^T Y = |X||Y|\cos\alpha \qquad \text{A2-2}$$

Figure A2.2: Illustration of Distance Measure from Inner Product

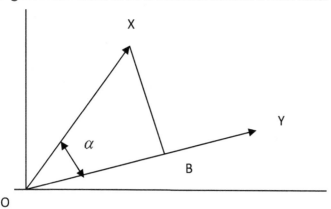

Now if we solve Equation A2-2 for the cosine of the angles between X and Y, we get the following formula.

$$A(X,Y) = \cos\alpha = \frac{X^T Y}{|X||Y|} = \frac{\sum_{i=1}^{i=n} x_i y_i}{\left(\left[\sum_{i=1}^{i=n} x_i^2\right]\left[\sum_{i=1}^{i=n} y_i^2\right]\right)^{\frac{1}{2}}} \qquad \text{A2-3}$$

The cosine of the angle between X and Y is a measure of *similarity* between X and Y. So how do these formulations become important? First, remember that when there are many different fields in your data set, they must be *transformed* onto a numeric scale that has the same meaning for all the fields being considered. Second, once in the *transformed* space, the distances between each of the records can be recorded using the preceding formulas, plus a few others that are not mentioned here; see Anderberg (1973) and Duda, Hart, and Stork (2001) for other formulas. Figure A2.2 can be visualized as having the points from your database as in the *transformed* space. Then, in order to understand the resulting distances, these *transformed* values can be brought back into their original dimensions and scales that have meaning to business users.

There have been many types of distance metrics created for special purposes. Distance metrics are especially suited for numeric data, while others are designed for use with certain types of data such as binary or categorical variables. There are dozens if not hundreds of published techniques for measuring the similarity of data records in a table or database. The basic classes of variables or fields in your database can be nicely put into the following four groups, although others groups or classes can and do exist as well:

- categorical (also called nominal)
- ordinal or ranks
- interval
- interval with an origin (also called ratio)

Categorical variables give us a classification system in which to place several unordered categories to which an item belongs. We can say that a store belongs to the northern region or the western region, but we cannot say that one is greater than the other is, or judge between the stores, only that they are located in these regions. In mathematical terms, we can say that $X \neq Y$ but not whether. Categorical measurements, then, denote that there is a difference in type between one or more levels versus another, but these measurements are not able to quantify that difference.

Ordinal or rank variables indicate to us that items have a certain specific order, but do not tell us how much difference there is between one item and another. Customers could be ranked as 1, 2, and 3, indicating that 3 is most valuable, 2 is next valuable, and 1 is least valuable, but only in relative terms. Ordinal measurements carry a lot more information than categorical measurements do. The ranking of categories should always be done subject to a particular condition. This is typically called *transitivity.* Transitivity means that if item A is ranked higher than B, and B higher than C, then A must be ranked higher than C. So, *A > B*, and *B > C*, then *A > C*.

Interval variables allow us to measure distances between two observations. If you were told that in Boston it is 48° and in southern New Hampshire it is 41°, then you would know that Boston is 7 degrees warmer than it is in New Hampshire. A special type of interval measurement is called a ratio scale. Ratios are not dimensional in nature; that is, they don't have a set of units that goes along with the measurement value. If a ratio is based from dividing the starting speed with an ending speed, then the units of speed (for example, miles per hour) cancel, and the unit is just a ratio without dimensions.

Intervals with an origin are what some call *true measure or ratio scale.* They are considered true because they have an origin as a proper reference system. Therefore, variables like age, weight, length, volume, and the like are *true* interval measures as they have a reference point of origin that is meaningful for comparison (Berry and Linoff 1997; Pyle 1999).

Geometric distance measures are well suited for interval variables and ones with an origin. When using categorical variables and rankings or ordinal measures, one needs to transform them into interval variables. This can be done in a variety of ways. There is a natural loss of information as one goes from interval with an origin, to ordinals like seniority, to categories like red or blue; there is a loss of information at each stage. This should be remembered when converting variables such as age into ranks.

In terms of mathematics, Table A2.4 gives some formal definitions of distance. Any function that takes two or more points and produces a single number describing a relationship between the points is a potential candidate for measure of association; however, a true distance metric must follow the rules in Table A2.4 (Berry and Linoff 1997).

Table A2.4: Distance Metrics Defined

$D(X,Y) = 0$ if and only if $X=Y$ ❶

$D(X,Y) \geq 0$ for all X and all Y ❷

$D(X,Y) = D(Y,X)$ ❸

$D(X,Y) \leq D(X,Z) + D(Z,Y)$ ❹

○ This property implies that if the distance is zero, then both points must be identical.
○ This property states that all distances must be positive. Vectors have both magnitude and direction; however, a distance between the vectors or points must be a positive number.
○ This property ensures symmetry by requiring the distance from X to Y to be the same as the distance from Y to X. In non-Euclidean geometry, this property does not necessarily hold true.

○ This property is known as the *triangle inequality,* and it requires that the length of one side of a triangle be no longer than the sum of the lengths of the other two sides (Anderberg 1973).

An example might be helpful to illustrate the above concepts. Table A2.5 shows seven customer sales records in a database. The fields are age of the person, revenue of items purchased, and state where the individual resides.

Computing the distance between rows 1 and 2 for age is straightforward. Distance (age)[1,2] = abs(55 − 28) = 27. However, if we want to compare this distance to rows 1 and 2 for revenue, Distance (revenue)[1,2] = abs(68 − 155) = 87 is not the same set of units. We need to transform these so that they are on the same relative scale; a scale between 0 and 1 would be one possible choice. We can do this by taking the absolute value of the difference and then dividing by the maximum difference. The maximum difference in age is the maximum − minimum; in age the max(age) is 55 and the min(age) is 26. Then, the normalized absolute value difference in age for rows 1 and 2 now is: Distance(age)[1,2] = abs(28 − 55) / (55 − 26) = 27 / 29 = 0.93103. The same type of computations can be done for revenue as well. Distance(revenue) [1,2] = abs(155 − 68) / (596 − 48) = 87 / 548 = 0.1587. State is a categorical variable and one method of transforming this is to transpose the state so that each unique level of state is a separate dummy variable for each level of state. This is shown in Table A2.6 below.

Table A2.5: Seven Customer Records in a Data Store

Row Number	Customer ID	Age (years)	Revenue ($)	State
1	372185321	28	$155	CA
2	075189457	55	$68	WA
3	538590043	32	$164	OH
4	112785896	40	$596	PA
5	678408574	26	$48	ME
6	009873687	45	$320	KS
7	138569322	37	$190	FL

Table A2.6: State Variable Dummy Transformations

State	CA	WA	OH	PA	ME	KS	FL
CA	1	0	0	0	0	0	0
WA	0	1	0	0	0	0	0
OH	0	0	1	0	0	0	0
PA	0	0	0	1	0	0	0
ME	0	0	0	0	1	0	0
KS	0	0	0	0	0	1	0
FL	0	0	0	0	0	0	1

Table A2.7: Normalized Distance Metrics of Age from Table A2.6

Customer ID	37218532	07518945	53859004	11278589	67840857	00987368	13856932
37218532	0	.					
07518945	4.60969	0	.				
53859004	3.76254	4.40064	0	.			
11278589	4.57006	4.90402	4.45998	0	.		
67840857	3.78974	4.704	3.83767	4.93738	0	.	
00987368	4.19027	4.09385	4.03964	4.04894	4.42432	0	.
13856932	3.8492	4.18897	3.77629	4.32954	3.96706	3.88526	0

This set of dummy variables allows the distances of a categorical variable like State and transforms it so that distances are measured on a scale between 0 and 1. These are not distances in miles between states, but likeness of records in the database to have a similar state name. If we now compute all the Euclidean distances and normalize them as shown earlier, Table A2.7 shows the matrix of normalized distances for the variable Age on the seven database records in Table A2.6.

For interval data, a general class of distance metrics for n-dimensional patterns is called the Minkowski metric and is expressed in the form of Equation A2-4.

$$D_p(X_j, X_k) = \left(\sum_{i=1}^{i=n} |X_{ij} - X_{ik}|^p \right)^{1/p}$$

A2-4

This metric is also known as the L_p norm. When p is 1, the metric is called the *city-block* or Manhattan distance, when p is 2, the Euclidean distance is obtained, and when p is 3, the Chebyshev metric is derived (Duda, Hart, and Stork 2001; Anderberg 1973). The distances in Table A2.7 were of the form when $p = 1$, and the values were normalized by dividing by the maximum less the minimum value. Many other derivations can be obtained in this fashion depending on the overall objective.

The k-Means Algorithm and Variants

The first to coin the term *k-means* was J. B. MacQueen (1967) who used this term to denote the process of assigning each data element to that cluster (of k clusters) with the nearest centroid (mean). The key part of the algorithm is that the cluster centroid is computed based on the cluster's current membership rather than its membership at the end of the last cycle of computations as other methods have done (Anderberg 1973). A *cluster* is nothing more than a group of database records that have something measurable in common; however, the basic structure of the groups is not known or defined. When a reference to a clustering algorithm is given, the reference is usually meaning an algorithm that is *undirected*.

Currently, the *k*-means method of cluster detection is one of the most widely used in practice. It also has quite a few variations. The *k*-means method was the main one that sparked the primary use in SAS. The selection of the number of clusters, *k*, has often been glossed over because the loop in the algorithm that selects a different *k* is really the analyst and not the computer program. What is typically done is after one selects a value of *k*, the resulting clusters are evaluated, then tried again with a different value of *k*. After each iteration, the strength of the resulting clusters can be evaluated by comparing the average distances between records in a cluster with the average distance between clusters, and there are other methods as well that are discussed later in this section. However, this type of iteration could be performed by the program, but an even more prominent issue arises in the cluster evaluation and that is the overall usefulness of the resulting clusters. Even well separated and clearly defined clusters, if not useful to the analyst or the desired application, have very little purpose in business or industry. The *k*-means algorithm is simple enough to specify (Duda, Hart, and Stork 2001).

k-Means Algorithm

Algorithm 1 (k-means clustering)

begin: initialize n, k and $u_1, u_2, u_3, \ldots u_k$

    classify n samples according to the nearest μ_j

      recompute μ_j

        until no change in μ_j

        return the values of $u_1, u_2, u_3, \ldots u_k$

  end:

where μ_j is the mean, n is the number of samples,

and k is the number of clusters.

Typically, the real value of *k* is sometimes denoted as *c*, and then the estimated value of *c* is denoted as *k*. By *real,* I mean the actual number of true clusters (if they exist) in the data set at hand, and *c* would be the estimate of *k*.

In geometry, all dimensions are equally important. As said earlier, what if certain fields in our database are measured in different units like that indicated in Table A2.5? These units must all be converted to the same *scale.* In Table A2.5, we cannot use one set of units such as dollars for revenue and try to convert age to dollars. The solution then is to map all the variables to a common *range* (like 0 to 1 or –1 to 1 or 0 to 100, and so on). That way, at least the ratios of change of one variable are comparable to the change in another variable. I refer this remapping to a common range as *scaling1*. The following list shows several methods for scaling variables to bring them into comparable ranges:

- Divide each variable by the mean (for example, each entry in a field is divided by the mean of the entire field).
- Subtract the mean value from each field and then divide by the standard deviation. In statistical terms, this is called a *z* score.
- Divide each field by the range (difference between the highest and lowest value) after subtracting the lowest value.
- Create a normal scale by the following Equation A2-5:

$$V_{norm} = \frac{V_i - \min(V_1...V_n)}{\max(V_1...V_n) - \min(V_1...V_n)}$$

A2-5

Variations of the k-Means Algorithm

The general form of the *k*-means clustering algorithm has a lot of variations. There are methods of selecting the initial seeds of the clusters, methods of computing the next centroid, or methods of using probability density rather than distance to associate records with the clusters. The *k*-means algorithm does have some drawbacks:

- It does not behave well when there are overlapping clusters.
- The cluster centers can be shifted due to outliers. Each record is either in or not in a cluster, although un-clustered outliers can be reviewed later. There is no notion of probability of cluster membership; for example, this record has an 80% likelihood of being in cluster 1.
- Fuzzy *k*-means clustering has been developed to simulate the third item in the preceding list of issues from classical *k*-means clustering. The *fuzzy* cluster membership is a probability that a database record belongs to a particular cluster or even to several clusters. The probability distribution often used is a Gaussian distribution (for example, a normal bell-shaped curve distribution). These variants of *k*-means are called Gaussian mixture models. Their name comes from a probability distribution assumed for highly dimensional types of problems. The seeds of the clusters are now the mean of a Gaussian distribution. During the estimation portion, this type of fuzzy membership is depicted in the key of Figure A2.3.

The darker cluster members have a probability of membership greater than 80%, the lighter cluster members have a probability of membership less than 80% but greater than 60%, and the lightest group of cluster members have less than 60% but greater than 40% probability of cluster membership. Although the lighter elements have a lower probability of being a member than the

Figure A2.3: Illustration of Fuzzy Cluster Membership

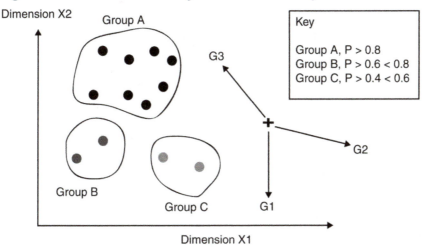

darker elements, they could have a high probability of being a member of another cluster. During the maximization step of the fuzzy algorithm, the association or responsibility that each Gaussian has for each data point will be used as weights immediately following the maximization step. Each Gaussian is shown as a G_1, G_2, G_n in Figure A2.3. Fuzzy clustering is sometimes referred to as *soft* clustering.

The Agglomerative Clustering Algorithm

There are diverse types of clustering and many algorithms to choose from a rather extensive list of possibilities. This section is intended to give you a flavor for the types of clustering methods. Most clustering techniques can be placed into two types, *disjoint* or *hierarchical*. Disjoint does not refer to bones that are out of socket; it refers to clusters that don't overlap, and each record in the data set belongs to only one cluster (or perhaps an outlier that does not belong to any particular cluster). Hierarchical clusters are ones in which a data record could belong to more than one cluster, and a hierarchical tree can be formulated that describes the clusters. This is particularly useful when building a taxonomy or trying to understand the possible structure in the data that might otherwise be unknown. Such a tree that shows this hierarchy is typically called a *dendrogram*, and an example of one is shown in Figure A2.4.

From the simple customer table in Table A2.5 and the coding of states as in Table A2.6, a simple clustering produces a simple dendrogram of hierarchical clusters as in Figure A2. The information about how far apart the clusters are from each other will be useful, as we consider how to measure the distance between clusters. In the first iteration through the cluster merge step, the clusters to be merged each contain only one entry or record so that the distance between clusters is the same as the distance between the records. However, on the second pass through and in subsequent passes, we need to update the similarity matrix with the distances from the

Figure A2.4: Hypothetical Example of Dendrogram Forming Hierarchical Clusters

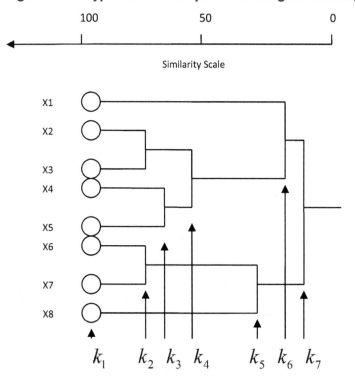

Figure A2.5: Hierarchical Clusters from Customer IDs in Table A2.5

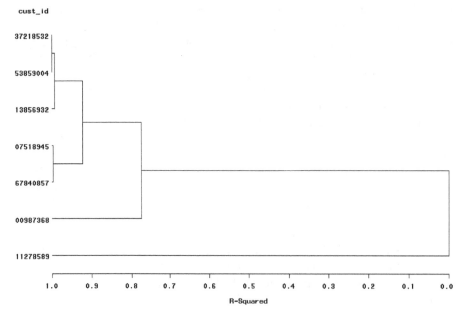

multi-record cluster to all of the other clusters. Again, there are choices to make on how we can measure the distances between the clusters. Here are the three most common approaches:

- Single linkage
- Complete linkage
- Difference or comparison of centroids

In the single linkage method, the distance between any of the clusters is determined by the distance between the closest members. This method produces clusters so that each member of a cluster is more closely related to each other than any other point outside that cluster; that is, it tends to find clusters that are more dense and closer to each other than in the other methods. In the complete linkage method, the distance between any of the clusters is given by the distance between their most distant members. This produces clusters with the property that all members lie within some known maximum distance of one another. Moreover, in the third method of centroids, the distance between the clusters is measured between the centroids of each (the centroids are the average or mean of the elements). Figure A2.6 shows a representation of the three methods (Berry and Linoff 1997).

For a complete survey of clustering methods, please refer to Xu and Tian (2015).

Figure A2.6: Three Common Methods for Measuring Cluster Distances

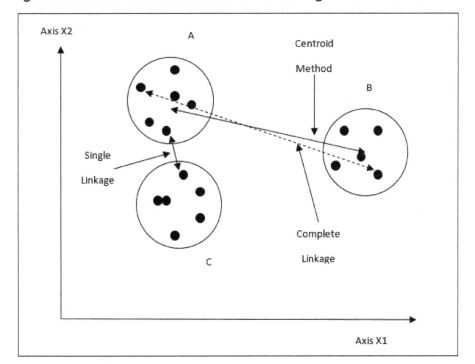

References

Anderberg, Michael R. 1973. Cluster Analysis for Applications. New York and London: Academic Press.

Berry, Michael J. A., and Gordon S. Linoff. 1997. Data Mining Techniques: for Marketing, Sales, and Customer Support. New York: Wiley.

Collica, Randall S. 2017. Customer Segmentation and Clustering Using SAS Enterprise Miner, 3rd ed., Cary, NC: SAS Institute, selected portions of chapter 3.

Duda, Richard O., Peter E. Hart, and David G. Stork. 2001. Pattern Classification. 2d ed. New York: Wiley.

MacQueen, James. B. 1967. "Some Methods for Classification and Analysis of Multivariate Observations." Proceedings of the Fifth Berkeley Symposium on Mathematical Statistics and Probability. Berkeley, CA: University of California Press. 1:281–297.

Pyle, Dorian. 1999. Data Preparation for Data Mining. San Francisco: Morgan Kaufmann Publishers.

Xu, Dongkuan and Yingjie Tian. 2015. "A Comprehensive Survey of Clustering Algorithms," Annals of Data Science, Ann. 2, pp. 165–193.

Ready to take your SAS® and JMP® skills up a notch?

Be among the first to know about new books, special events, and exclusive discounts.
support.sas.com/newbooks

Share your expertise. Write a book with SAS.
support.sas.com/publish

Continue your skills development with free online learning.
www.sas.com/free-training

sas.com/books
for additional books and resources.

THE POWER TO KNOW®

www.ingramcontent.com/pod-product-compliance
Lightning Source LLC
La Vergne TN
LVHW080117070326
832902LV00015B/2645